IN THE MIND OF A MAINTENANCE MAN

IN THE MIND
OF A
MAINTENANCE MAN

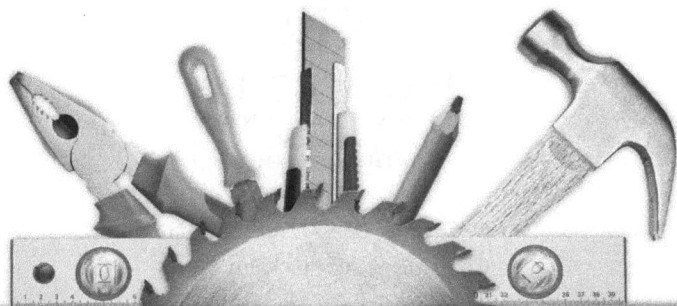

KEITH TURNER

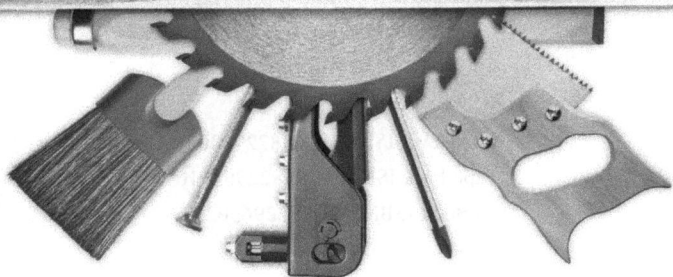

PALMETTO
P U B L I S H I N G
Charleston, SC
www.PalmettoPublishing.com

In The Mind Of A Maintenance Man
Copyright © 2024 by Keith Turner

First Edition

Hardcover ISBN: 9798822963009
Paperback ISBN: 9798822963016
eBook ISBN: 9798822963023

TABLE OF CONTENTS

HVAC

FROZEN AC CONDENSER UNIT

A frozen AC condenser unit typically indicates a problem with the system's airflow or refrigerant levels. The most common issues associated with a frozen AC condenser unit include:

Restricted Airflow: Poor airflow across the evaporator coil can cause it to become too cold, leading to condensation freezing on the coil and eventually extending to the condenser unit. Restricted airflow can result from dirty air filters, blocked vents, closed registers, or obstructed ductwork.

Low Refrigerant Levels: Insufficient refrigerant in the system can cause the evaporator coil to become too cold, leading to ice formation. Low refrigerant levels can result from leaks in the refrigerant lines or improper installation.

Faulty Thermostat: A malfunctioning thermostat may cause the AC system to run longer than necessary, resulting in excessively low temperatures at the evaporator coil and potential freezing.

Dirty Coils: Accumulation of dirt, dust, or debris on the evaporator coil or condenser coil can impair heat transfer, causing the system to run inefficiently and potentially leading to freezing.

Defective Blower Motor: A malfunctioning blower motor can result in inadequate airflow across the evaporator coil, causing it to become too cold and freeze.

Blocked Drainage: If the condensate drain line is blocked or clogged, water may accumulate and freeze on the evaporator coil, eventually extending to the condenser.

Repairing a frozen AC condenser unit involves several steps, and it's essential to address the underlying cause of the freezing to prevent future occurrences.
Here's a general guide:

Turn Off the AC: Switch off the air conditioning system to prevent further ice buildup and potential damage to the unit.

Thaw the System: Allow the frozen components, such as the evaporator coil and condenser unit, to thaw completely. This may take several hours or longer depending on the extent of the ice buildup.

Inspect Air Filters: Check the air filters for dirt, dust, or debris. If they are dirty or clogged, replace them with clean filters to ensure proper airflow.

Check Thermostat Settings: Verify that the thermostat is set to the correct temperature and operating mode (cooling). Ensure that the thermostat is functioning correctly and is accurately reading the temperature.

Inspect Refrigerant Levels: If you suspect low refrigerant levels, contact a qualified HVAC (heating, ventilation, and air conditioning) technician to perform a refrigerant leak test and recharge the system if necessary. Refrigerant leaks should be repaired promptly to prevent further issues.

Clean Coils and Fins: Remove any dirt, dust, or debris from the evaporator coil and condenser coil using a soft brush or vacuum cleaner. Be gentle to avoid damaging the fins or coil.

Check Blower Motor: Verify that the blower motor is functioning correctly and providing adequate airflow across the evaporator coil. If the motor is faulty, it may need to be repaired or replaced by a professional.

Inspect Drainage System: Check the condensate drain line for blockages or clogs. Clear any obstructions to ensure proper drainage and prevent water buildup.

Monitor System Operation: After completing repairs and maintenance, turn on the AC system and monitor its operation. Ensure that the system is cooling properly without any signs of ice formation or reduced airflow.

Schedule Regular Maintenance: To prevent future issues, schedule regular maintenance for your AC system, including cleaning coils, replacing air filters, and inspecting for any potential problems.

CONDENSER FAN NOT WORKING

When the condenser fan is not working, several common problems could be causing the issue. Here are some of the most frequent ones:

Faulty Motor: The fan motor may have failed due to wear and tear, electrical issues, or overheating. This can prevent the fan from spinning and cooling the condenser coil.

Capacitor Issues: The capacitor provides the initial boost of electrical energy to start the fan motor. If the capacitor is faulty or has failed, the fan may not start or run properly.

Electrical Problems: Faulty wiring, loose connections, or tripped circuit breakers can interrupt the flow of electricity to the fan motor, preventing it from functioning correctly.

Blocked or Obstructed Fan: Debris such as leaves, dirt, or other objects may have accumulated around the fan blades, preventing them from turning. This can cause the motor to overheat and fail.

Motor Overload: If the condenser unit is working under heavy load or in extremely hot conditions, the motor may become overloaded and shut down to prevent damage.

Thermostat Malfunction: A malfunctioning thermostat may not be signaling the condenser fan to turn on when needed, resulting in inadequate cooling.

Defective Relay or Contactor: The relay or contactor controls the flow of electricity to the condenser fan motor. If it is faulty or damaged, the fan may not receive power.

Fan Blade Damage: If the fan blades are bent, warped, or damaged, they may not be able to spin properly, resulting in reduced airflow and cooling efficiency.

Low Voltage Supply: Insufficient voltage reaching the condenser unit can prevent the fan motor from operating at full capacity or starting at all.

Motor Bearings: Over time, the bearings in the fan motor can wear out, causing the motor to become noisy or seize up altogether.

Repairing a condenser fan that is not working involves identifying and addressing the underlying cause of the problem. Here's a step-by-step guide for repairing a condenser fan:

Turn Off Power: Before performing any repairs, turn off the power to the condenser unit at the circuit breaker or disconnect switch to prevent electrical shock.

Inspect Fan Blades: Check the fan blades for any signs of damage, such as bending, warping, or debris accumulation. Clean the blades and remove any obstructions that may be preventing them from spinning freely.

Check Electrical Connections: Inspect the wiring connections at the fan motor, capacitor, relay, and contactor for any signs of damage, corrosion, or loose connections. Tighten any loose connections and replace any damaged wiring as needed.

Test Capacitor: Use a multimeter to test the capacitor for proper functionality. If the capacitor is faulty or has failed, replace it with a new one of the same rating.

Test Voltage Supply: Verify that the condenser unit is receiving the correct voltage supply. If the voltage is low, consult a qualified C to address the issue.

Check Relay or Contactor: Test the relay or contactor to ensure that it is sending power to the fan motor when the thermostat calls for cooling. If the relay or contactor is faulty, replace it with a new one.

Inspect Motor Bearings: Check the motor bearings for signs of wear or damage. If the bearings are worn out, lubricate them with oil if possible. If the bearings are severely worn or damaged, replace the fan motor.

Test Motor Windings: Use a multimeter to test the motor windings for continuity. If any of the windings are open or shorted, replace the fan motor.

Reset Overload Protection: If the fan motor has overload protection, reset it according to the manufacturer's instructions.

Turn On Power: After completing the repairs, turn on the power to the condenser unit and test the fan operation. Verify that the fan blades are spinning freely and that the motor is running smoothly.

AC COMPRESSOR NOT COMING ON

If your AC compressor is not coming on, several common reasons could be causing the issue. Here are some potential causes to consider:

Thermostat Settings: Ensure that the thermostat is set to "Cool" mode and that the temperature is set lower than the current room temperature. If the thermostat is set incorrectly, it may not signal the AC compressor to turn on.

Tripped Breaker or Blown Fuse: Check the circuit breaker or fuse box to see if the circuit that powers the AC compressor has tripped or if the fuse has blown. Reset the breaker or replace the fuse if necessary.

Faulty Thermostat: A malfunctioning thermostat may not be sending the correct signals to the AC system, preventing the compressor from turning on. Test the thermostat by setting it to a lower temperature and listening for a click indicating that it's sending a signal to the system.

Faulty Capacitor: The capacitor provides the initial boost of electrical energy to start the compressor. If the capacitor is faulty or has failed, the compressor may not start. Inspect the capacitor for signs of bulging, leaking, or damage and replace it if necessary.

Low Refrigerant Levels: Insufficient refrigerant in the system can cause the low-pressure switch to prevent the compressor from turning on to protect it from damage. Have a qualified HVAC technician check for refrigerant leaks and recharge the system if necessary.

Faulty Contactor: The contactor is an electrical switch that controls the flow of electricity to the compressor. If the contactor is faulty or worn out, it may not make proper contact, preventing the compressor from turning on. Inspect the contactor for signs of burning or pitting and replace it if necessary.

Tripped Pressure Switch: The pressure switches in the refrigerant lines monitor the system's pressure levels. If the pressure switch detects abnormal pressures, it may prevent the compressor from turning on. Have a qualified technician check the pressure switches and reset them if necessary.

Faulty Compressor: If the compressor itself is faulty or damaged, it may not be able to start or operate properly. Signs of a faulty compressor include loud noises, excessive vibration, or visible damage. If the compressor is determined to be the issue, it will likely need to be replaced.

Wiring Issues: Inspect the wiring connections to the compressor for any signs of damage, corrosion, or loose connections. Ensure that all wires are securely connected and that there are no breaks in the wiring.

Safety Switches: Some AC systems have safety switches that can prevent the compressor from turning on if certain conditions are not met, such as a clogged condensate drain or a malfunctioning fan motor. Check for any safety switches that may be tripped and address the underlying issue

Repairing an AC compressor that is not coming on involves diagnosing the underlying cause of the issue and then addressing it accordingly. Here's a step-by-step guide to repairing an AC compressor:

Turn Off Power: Before beginning any repairs, turn off the power to the AC system at the circuit breaker or disconnect switch to prevent electrical shock.

Check Thermostat Settings: Ensure that the thermostat is set to "Cool" mode and that the temperature setting is lower than the current room temperature. Adjust the thermostat if necessary.

Inspect Breaker/Fuse: Check the circuit breaker or fuse box to see if the circuit that powers the AC compressor has tripped or if the fuse has blown. Reset the breaker or replace the fuse if necessary.

Test Thermostat: Test the thermostat by setting it to a lower temperature and listening for a click indicating that it's sending a signal to the system. If the thermostat is faulty, replace it with a new one.

Inspect Capacitor: Inspect the capacitor for signs of bulging, leaking, or damage. Use a multimeter to test the capacitor for proper functionality. If the capacitor is faulty, replace it with a new one of the same rating.

Check Refrigerant Levels: Have a qualified HVAC technician check for refrigerant leaks and recharge the system if necessary. Low refrigerant levels can prevent the compressor from turning on.

Inspect Contactor: Inspect the contactor for signs of burning, pitting, or wear. If the contactor is faulty, replace it with a new one.

Reset Pressure Switches: If the pressure switches have tripped, reset them according to the manufacturer's instructions.

Inspect Wiring: Inspect the wiring connections to the compressor for any signs of damage, corrosion, or loose connections. Ensure that all wires are securely connected and that there are no breaks in the wiring.

Test Compressor: If all other components check out fine, test the compressor itself to determine if it's functioning properly. This typically requires specialized equipment and should be performed by a qualified HVAC technician.

Address Safety Switches: Check for any safety switches that may be tripped and address the underlying issue causing the switch to trip, such as a clogged condensate drain or a malfunctioning fan motor.

Turn On Power: After completing the repairs, turn on the power to the AC system and test the compressor operation. Monitor the system for any signs of abnormal operation or issues.

CONTACTOR NOT PULLING IN

If your AC condenser contactor is not pulling in, preventing the compressor from turning on, several common reasons could be causing the issue. Here are some potential causes to consider:

No Power Supply: Ensure that there is power reaching the contactor. Check the circuit breaker or fuse box to make sure the circuit supplying power to the condenser unit is not tripped or blown. If it is, reset the breaker or replace the fuse.

Faulty Contactor: Examine the contactor for signs of damage, wear, or burning. If the contacts are pitted or burned, they may not be making proper contact, preventing the contactor from pulling in. Replace the contactor if necessary.

Low Control Voltage: Use a multimeter to test the control voltage at the contactor coil terminals. The control voltage is typically 24 volts AC. If there is no voltage present, check for a faulty thermostat, control board, or transformer that may not be sending the signal to the contactor.

Faulty Thermostat or Control Board: A malfunctioning thermostat or control board may not be sending the signal to the contactor to engage. Test the thermostat and control board for proper operation and replace them if necessary.

Wiring Issues: Inspect the wiring connections to the contactor for any signs of damage, corrosion, or loose connections. Ensure that all wires are securely connected and that there are no breaks in the wiring.

Faulty Safety Switches: Some AC systems have safety switches that can prevent the contactor from pulling in if certain conditions are not met, such as a clogged condensate drain or a malfunctioning fan motor. Check for any safety switches that may be tripped and address the underlying issue.

Pressure Switches: Check the pressure switches in the refrigerant lines. If abnormal pressure levels are detected, the pressure switches may prevent the contactor from engaging. Reset the pressure switches if necessary.

Control Transformer: Test the control transformer that supplies power to the thermostat and control circuitry. If the transformer is faulty or not providing the correct voltage, it may prevent the contactor from pulling in. Replace the transformer if necessary.

Faulty Low-Voltage Wiring: Inspect the low-voltage wiring from the thermostat to the contactor for any damage or breaks. Repair or replace any damaged wiring.

Compressor Overload: If the compressor has overheated or is overloaded, it may have a safety feature that prevents the contactor from engaging until the compressor has cooled down. Allow the compressor to cool and reset any overload protection if necessary.

Fixing a condenser contactor that is not pulling in involves identifying the underlying cause of the issue and then addressing it accordingly. Here's a step-by-step guide to fixing the problem:

Turn Off Power: Before beginning any repairs, turn off the power to the AC system at the circuit breaker or disconnect switch to prevent electrical shock.

Inspect the Contactor: Examine the contactor for signs of damage, wear, or burning. If the contacts are pitted or burned, they may not be making proper contact, preventing the contactor from pulling in. Replace the contactor if necessary.

Check Power Supply: Ensure that there is power reaching the contactor. Check the circuit breaker or fuse box to make sure the circuit supplying power to the condenser unit is not tripped or blown. If it is, reset the breaker or replace the fuse.

Test Control Voltage: Use a multimeter to test the control voltage at the contactor coil terminals. The control voltage is typically 24 volts AC. If there is no voltage present, check for a faulty thermostat, control board, or transformer that may not be sending the signal to the contactor.

Inspect Wiring Connections: Inspect the wiring connections to the contactor for any signs of damage, corrosion, or loose connections. Ensure that all wires are securely connected and that there are no breaks in the wiring.

Reset Safety Switches: Some AC systems have safety switches that can prevent the contactor from pulling in if certain conditions are not met, such as a clogged condensate drain or a malfunctioning fan motor. Reset any safety switches that may be tripped and address the underlying issue.

Replace Faulty Components: Replace any faulty components that are identified during the inspection, such as the thermostat, control board, transformer, or pressure switches.

Test the System: After completing the repairs, turn on the power to the AC system and test the operation. Monitor the system to ensure that the contactor is pulling in properly and that the compressor is turning on as expected.

Regular Maintenance: To prevent future issues, schedule regular maintenance for your AC system, including inspecting and cleaning the contactor, checking wiring connections, and testing electrical components.

3-AMP FUSE KEEPS POPPING

If your HVAC system's 3-amp fuse keeps popping, it could indicate several common problems:

Overloaded Circuit: Check if the circuit is overloaded. If the HVAC system shares the circuit with other high-power appliances, it might be drawing too much current, causing the fuse to blow.

Short Circuit: A short circuit in the HVAC system's wiring or components can cause the fuse to blow repeatedly. Inspect the wiring for any signs of damage or exposed wires.

Faulty Capacitor: A faulty capacitor in the HVAC system can cause excessive current draw, leading to fuse blowing. Capacitors store electrical energy and regulate the flow of electricity to the motors. If they fail, they can cause the system to draw more current than normal.

Faulty Motor: A malfunctioning blower motor or compressor motor can draw too much current, causing the fuse to blow. Motors can fail due to age, wear and tear, or lack of maintenance.

Dirty Air Filters or Coils: Restricted airflow due to dirty air filters or coils can cause the system to work harder, leading to increased current draw and blown fuses.

Refrigerant Leaks: Low refrigerant levels can cause the compressor to overheat and draw excessive current, leading to blown fuses.

Improper Sizing: If the HVAC system is improperly sized for the space it's supposed to cool or heat, it may run excessively, causing fuses to blow.

Faulty Control Board: The control board in the HVAC system may be malfunctioning, causing erratic behavior and excessive current draw.

Repairing a constantly blowing 3-amp fuse in an HVAC system can be challenging and may require some electrical and HVAC expertise. Here's a general guide on how you might approach repairing it:

Turn Off Power: Before attempting any repairs, make sure to turn off the power to the HVAC system at the circuit breaker or fuse box to prevent electric shock.

Inspect Wiring: Carefully inspect the wiring throughout the HVAC system for any signs of damage, such as frayed or exposed wires. If you find any issues, repair or replace the wiring as necessary.

Check Components: Check all components of the HVAC system, including the compressor, blower motor, fan motor, and control board, for any signs of damage or malfunction. Look for burnt or overheated components.

Test Capacitors: Capacitors can often be the culprit behind blown fuses. Use a multimeter to test the capacitors for proper functionality. If any capacitors are faulty, replace them with compatible replacements.

Inspect Motors: Test the blower motor and compressor motor for proper operation. If either motor is drawing excessive current or showing signs of failure, it may need to be repaired or replaced.

Check Refrigerant Levels: If the HVAC system uses a refrigerant, check the refrigerant levels to ensure they're within the recommended range. Low refrigerant levels can cause the compressor to overheat and draw excessive current.

Clean Air Filters and Coils: If dirty air filters or coils are restricting airflow, clean or replace them as necessary. Restricted airflow can cause the system to work harder and draw more current.

Verify Sizing: Ensure that the HVAC system is properly sized for the space it's intended to cool or heat. An improperly sized system may run excessively, leading to blown fuses.

Inspect Control Board: Check the control board for any signs of damage or malfunction. If the control board is faulty, it may need to be replaced.

Replace Fuse: After addressing any issues you find, replace the blown fuse with a new one of the correct amperage rating.

Turn On Power: Once all repairs are complete, turn the power back on to the HVAC system and monitor its operation.

GAS FURNACE PILOT LIGHT WILL NOT STAY LIT

When a gas furnace pilot light won't stay lit, it typically indicates one of several common issues:

Thermocouple Issues: The thermocouple is a safety device that detects the presence of the pilot flame. If it's malfunctioning or positioned incorrectly, it may not generate enough voltage to keep the gas valve open, causing the pilot light to extinguish. Cleaning or replacing the thermocouple often resolves this issue.

Dirty Pilot Orifice: Dirt, dust, or debris can clog the pilot orifice, disrupting the flow of gas to the pilot light. Cleaning the orifice with compressed air or a thin wire can often solve the problem.

Weak Pilot Flame: A pilot flame that is too weak may not generate enough heat to keep the thermocouple functioning properly. Adjusting the pilot flame height according to the manufacturer's instructions can sometimes resolve this issue.

Gas Supply Issues: Low gas pressure or a disrupted gas supply can cause the pilot light to go out. Check the gas valve to ensure it's fully open and that there are no leaks in the gas line. If you suspect a gas supply issue, contact your gas utility provider for assistance.

Faulty Gas Valve: A faulty gas valve may not maintain the proper gas flow to the pilot light, causing it to go out. If other troubleshooting steps fail to resolve the issue, the gas valve may need to be replaced.

Drafts or Ventilation Issues: Strong drafts or poor ventilation near the furnace can extinguish the pilot light. Ensure that vents and flues are clear of obstructions and that there are no drafts near the appliance.

Thermostat Problems: If the thermostat is faulty or set incorrectly, it may not call for heat, causing the pilot light to go out. Check the thermostat settings and replace the thermostat if necessary.

Electrode or Igniter Issues: For furnaces with electronic ignition systems, problems with the electrode or igniter can prevent the pilot light from staying lit. Ensure that the electrode or igniter is clean and positioned correctly.

To repair a gas furnace pilot light that won't stay lit, you can follow these steps:

Turn Off Gas and Electricity: Before starting any repair work, turn off the gas supply to the furnace at the main gas valve and switch off the electricity to the furnace at the breaker box. This ensures your safety while working on the appliance.

Clean the Pilot Orifice and Burner: Use compressed air or a small brush to clean the pilot orifice and burner. Dust, dirt, or debris buildup can obstruct the flow of gas, leading to a weak or extinguished pilot light.

Check the Thermocouple: Inspect the thermocouple for proper positioning and cleanliness. If it's dirty, clean it gently with fine sandpaper or replace it if it's damaged or malfunctioning. Make sure the thermocouple is in the path of the pilot flame.

Adjust the Pilot Flame: Ensure the pilot flame is steady and blue. If it's flickering or yellow, it may be too weak. Using the pilot light adjustment screw, adjust the pilot flame height according to the manufacturer's instructions.

Check Gas Supply: Verify that the gas supply valve is fully open. If it's partially closed or obstructed, it can affect the pilot light's stability. Also check for any leaks in the gas line.

Inspect for Drafts: Look for any drafts near the furnace that could be blowing out the pilot light. Seal any gaps or cracks around doors, windows, or vents to prevent drafts from affecting the flame.

Verify Ignition System: For furnaces with electronic ignition systems, check the electrode or igniter for proper positioning and cleanliness. Ensure that the igniter sparks consistently when the furnace calls for heat.

> **If you've checked all the above steps and the pilot light still won't stay lit, you may need to replace faulty components such as the thermocouple, gas valve, or electronic ignition module.**

Test the Furnace: After completing the repairs, turn on the gas supply and electricity to the furnace. Test the furnace to ensure the pilot light stays lit and the system operates properly.

FURNACE FAN SHORT CYCLING

Furnace fan short cycling, where the fan turns on and off frequently, can be caused by several common issues:

Dirty Air Filter: A dirty or clogged air filter restricts airflow, causing the furnace to overheat. When the furnace overheats, the limit switch shuts off the burner to prevent damage. Once the furnace cools down, the fan may turn on again, leading to short cycling. Replace the air filter regularly to ensure proper airflow.

Incorrect Thermostat Settings: Check the thermostat settings to ensure they're correctly configured. If the thermostat is set too close to the desired temperature, or if the heat anticipator is set incorrectly, it can cause the furnace to short cycle.

Thermostat Location: The thermostat's location plays a significant role in its accuracy. If the thermostat is installed in a drafty area or near heat sources like lamps or sunlight, it may register incorrect temperatures, leading to short cycling.

Blocked Vents or Registers: Blocked vents or registers restrict airflow, causing the furnace to overheat and short cycle. Ensure that all vents and registers are open and unobstructed to allow proper airflow throughout the house.

Dirty or Faulty Components: Dirty blower wheels, clogged evaporator coils, or malfunctioning fan motors can cause the furnace to overheat and short cycle. Regular maintenance, including cleaning and lubricating moving parts, can help prevent these issues.

Improper Ductwork Design: Inadequate or poorly designed ductwork can lead to airflow restrictions and uneven heating, causing the furnace to short cycle. Consult with an HVAC professional to evaluate and, if necessary, redesign the ductwork for better efficiency.

Faulty Limit Switch: A malfunctioning limit switch may inaccurately detect the furnace's temperature, causing it to shut off prematurely and short cycle. If the limit switch is faulty, it may need to be replaced by a qualified technician.

Oversized Furnace: If the furnace is too large for the space it's heating, it may heat the home too quickly, causing it to cycle on and off frequently. Consult with an HVAC professional to determine if the furnace is properly sized for your home.

Electrical Issues: Loose electrical connections, faulty relays, or issues with the control board can cause erratic furnace operation, including short cycling. A qualified technician can diagnose and repair any electrical issues.

To repair furnace fan short cycling, you can follow these steps:

Check and Replace Air Filter: Start by checking the air filter. If it's dirty or clogged, replace it with a new one. A dirty filter restricts airflow, causing the furnace to overheat and short cycle.

Adjust Thermostat Settings: Ensure the thermostat is set correctly. Make sure it's not set too close to the desired temperature, as this can cause the furnace to turn on and off frequently. Additionally, check the heat anticipator setting if your thermostat has one, and adjust it according to the manufacturer's instructions.

Check Thermostat Location: Verify that the thermostat is installed in a suitable location away from drafts, heat sources, or direct sunlight, which can affect its accuracy. If needed, relocate the thermostat to a more appropriate location.

Clear Vents and Registers: Ensure that all vents and registers throughout the house are open and unobstructed to allow proper airflow. Move furniture or objects blocking the vents and registers if necessary.

Inspect and Clean Components: Check the blower wheel, evaporator coil, and fan motor for dirt, debris, or obstructions. Clean these components as needed to ensure proper airflow and operation.

Check Ductwork: Inspect the ductwork for any blockages, leaks, or restrictions that could impede airflow. Repair or seal any ductwork issues to improve efficiency.

Test Limit Switch: Test the limit switch to ensure it's functioning correctly. If the switch is faulty or inaccurate, it may need to be replaced by a qualified technician.

Verify Electrical Connections: Check for loose electrical connections, damaged wires, or faulty relays in the furnace's electrical system. Tighten connections and replace any damaged components as needed.

Verify Furnace Sizing: Ensure that the furnace is properly sized for the space it's heating. An oversized furnace can cause short cycling. Consult with an HVAC professional to determine if the furnace size is appropriate for your home.

Reset Control Board: If your furnace has a control board, reset it by turning off the power to the furnace for a few minutes, then turning it back on. This can sometimes resolve issues with erratic furnace operation.

EXCESSIVE CONDENSATION

Excessive condensation in an HVAC system can indicate several common problems:

Poor Insulation: Inadequate insulation around ductwork or on refrigerant lines can lead to condensation forming on these surfaces. Insulating these components properly can help prevent condensation buildup.

Improper Sizing: Oversized HVAC systems can cool the air too quickly, leading to excess condensation on the evaporator coil. This can occur when the system cycles on and off frequently without running long enough to remove humidity effectively.

Clogged Drain Line: A clogged or obstructed condensate drain line can cause water to back up and overflow, leading to excess moisture in the system. Regularly inspect and clean the condensate drain line to prevent blockages.

Dirty Air Filters: Dirty or clogged air filters restrict airflow, causing the evaporator coil to become colder than usual. This can result in excessive condensation forming on the coil. Replace air filters regularly to maintain proper airflow and prevent condensation issues.

Low Refrigerant Levels: Low refrigerant levels can cause the evaporator coil to freeze up, leading to excess condensation when the ice melts. Have a qualified HVAC technician inspect the system for refrigerant leaks and recharge the refrigerant if necessary.

Leaky Ductwork: Leaks or gaps in the ductwork can allow warm, humid air to infiltrate the system, leading to condensation forming on cold surfaces. Seal any duct leaks to prevent excess moisture from entering the system.

Faulty Humidifier: If your HVAC system includes a humidifier, a malfunctioning humidistat or faulty water supply valve can lead to excess humidity in the system. Check and adjust the humidistat settings as needed, and repair or replace any faulty components.

High Indoor Humidity Levels: Excessive indoor humidity can contribute to condensation issues in the HVAC system. Use a dehumidifier to reduce indoor humidity levels, especially in areas with poor ventilation or high moisture sources (such as bathrooms or kitchens).

Inadequate Ventilation: Poor ventilation in the home can trap moisture indoors, leading to condensation problems in the HVAC system. Ensure proper ventilation by using exhaust fans in bathrooms and kitchens and opening windows when the weather permits.

Warm Air Contacts Cold Surfaces: Condensation can occur when warm air comes into contact with cold surfaces, such as uninsulated ducts or air conditioning components. Properly insulate ductwork and equipment to minimize temperature differentials and reduce condensation.

To repair excessive condensation in your HVAC system, you can follow these steps:

Inspect and Clean Air Filters: Start by checking the air filters. If they're dirty or clogged, replace them with new ones. Dirty filters restrict airflow, leading to issues like freezing coils and excess condensation.

Check and Clean Condensate Drain Line: Inspect the condensate drain line for clogs or obstructions. Use a wet/dry vacuum or a pipe brush to remove any buildup and ensure proper drainage.

Check Insulation: Inspect the insulation around ductwork and refrigerant lines. Proper insulation helps prevent condensation from forming on these surfaces. Add or replace insulation as needed.

Adjust Humidifier Settings: If your HVAC system includes a humidifier, check the settings and adjust them if necessary. Lowering the humidity level can reduce excess moisture in the system.

Seal Duct Leaks: Inspect the ductwork for leaks or gaps and seal them using foil tape or mastic sealant. Leaky ducts can allow warm, humid air to enter the system, contributing to condensation issues.

Check Refrigerant Levels: Have a qualified HVAC technician inspect the refrigerant levels in your system. Low refrigerant levels can cause the evaporator coil to freeze up, leading to excess condensation when it thaws.

Clean Evaporator Coil: Inspect the evaporator coil for dirt or debris buildup. Clean the coil using a soft brush or a commercial coil cleaner to ensure proper heat transfer and prevent condensation issues.

Improve Ventilation: Ensure proper ventilation in your home by using exhaust fans in bathrooms and kitchens and opening windows when the weather permits. Adequate ventilation helps remove excess moisture from the air.

Install a Dehumidifier: Consider installing a whole-house dehumidifier to help control indoor humidity levels. A dehumidifier removes excess moisture from the air, reducing the likelihood of condensation forming in your HVAC system.

HEAT STRIP NOT WORKING

When the heat strip in an HVAC air handler isn't working, it could be due to several common problems:

Faulty Heating Element: The heating element itself may be defective or damaged, preventing it from generating heat. Over time, heating elements can wear out and require replacement.

Tripped Circuit Breaker: Check the circuit breaker or fuse box to see if the circuit supplying power to the heat strip has tripped. If so, reset the breaker or replace the fuse. A tripped breaker could indicate an electrical issue or overload.

Thermostat: A malfunctioning thermostat may not be signaling the heat strip to turn on. Check the thermostat settings and ensure it's set to heat mode, with the desired temperature properly configured.

Faulty Relay or Sequencer: The heat strip is typically controlled by relays or sequencers that turn it on and off. If these components are faulty, they may not be sending the correct signals to activate the heat strip.

Broken Wiring or Connections: Inspect the wiring and connections leading to the heat strip for any signs of damage, corrosion, or disconnection. Faulty wiring can prevent the heat strip from receiving power or operating correctly.

High-Limit Switch Tripped: The heat strip may have a high-limit switch that shuts it off if it overheats. If the switch is tripped, it needs to be reset or replaced. Overheating can occur due to restricted airflow or other issues.

Dirty Air Filter or Coils: A clogged air filter or dirty evaporator or heating coils can restrict airflow, causing the heat strip to overheat and shut off prematurely. Replace or clean the air filter and coils regularly to prevent this issue.

Incorrect voltage or wiring:

Configuration: Ensure that the heat strip is receiving the correct voltage and that the wiring is installed according to manufacturer specifications. Incorrect voltage or wiring configuration can prevent the heat strip from functioning properly.

Reversing Valve Issues (Heat Pumps Only): If the HVAC system is a heat pump, problems with the reversing valve can prevent the heat strip from operating correctly. The reversing valve may need to be inspected and repaired by a qualified technician.

Control Board Malfunction: A malfunctioning control board can prevent the heat strip from receiving the correct signals to turn on. If other troubleshooting steps don't identify the issue, the control board may need to be replaced.

Repairing a malfunctioning heat strip in an HVAC air handler involves several steps, depending on the specific problem. Here's a general guide:

Turn Off Power: Before beginning any repairs, turn off the power to the HVAC system at the circuit breaker or disconnect switch to prevent electric shock.

Inspect Heating Element: Check the heating element for any visible signs of damage or wear, such as broken coils or discoloration. If the heating element appears damaged, it will likely need to be replaced.

Check Circuit Breaker: Verify that the circuit breaker or fuse supplying power to the heat strip is not tripped. If it has tripped, reset the breaker or replace the fuse. If it continues to trip, there may be an electrical issue that needs to be addressed.

Test Thermostat: Ensure the thermostat is set to heat mode and the temperature is set above room temperature. Test the thermostat by raising the set temperature to see if it activates the heat strip.

Inspect Relays or Sequencers: If the heat strip is controlled by relays or sequencers, check these components for proper operation. Test each relay or sequencer to ensure it's sending the correct signals to activate the heat strip.

Check Wiring and Connections: Inspect the wiring and connections leading to the heat strip for any signs of damage, corrosion, or loose connections. Repair or replace any damaged wiring or connectors.

Reset High-Limit Switch: If the heat strip has a high-limit switch that has tripped due to overheating, reset the switch according to the manufacturer's instructions. Address any factors contributing to overheating, such as restricted airflow or dirty coils.

Clean Air Filter and Coils: Replace or clean the air filter and evaporator coils to ensure proper airflow. Restricted airflow can cause the heat strip to overheat and malfunction.

Verify Voltage and Wiring Configuration: Ensure that the heat strip is receiving the correct voltage and that the wiring is installed according to manufacturer specifications. Incorrect voltage or wiring configuration can prevent the heat strip from operating properly.

Test Control Board: If other troubleshooting steps do not identify the issue, the control board may be malfunctioning. Test the control board for proper operation or replace it if necessary.

REFRIGERATION LEAKS

Refrigeration leaks in HVAC systems can stem from various causes, leading to a loss of cooling capacity and potentially harmful environmental impacts due to refrigerant emissions. Some common problems associated with refrigeration leaks include:

Corrosion: Over time, the copper tubing used in HVAC systems can corrode due to chemical reactions with the refrigerant or moisture in the system. Corrosion weakens the tubing, making it prone to developing leaks.

Vibration Damage: Vibrations from the operation of HVAC equipment can cause joints, connections, and tubing to loosen or wear down over time, leading to refrigerant leaks.

Poor Installation: Improper installation, including inadequate brazing or soldering of tubing joints, can create weak points in the system where leaks are more likely to occur. It's crucial to have HVAC systems installed by qualified technicians using proper techniques and materials.

Physical Damage: Accidental damage from impacts, construction work, or other external factors can puncture or rupture refrigerant lines, causing leaks.

Manufacturing Defects: Defects in HVAC system components, such as tubing, coils, valves, or fittings, can lead to refrigerant leaks. While less common, manufacturing defects can occur and may not become apparent until the system is in operation.

Pressure Fluctuations: Fluctuations in system pressure, such as those caused by overcharging, undercharging, or improper operation, can stress tubing and connections, leading to leaks.

Freezing and Thawing Cycles: In systems that experience freezing and thawing cycles, such as those in cold climates or where equipment operates intermittently, the expansion and contraction of components can cause fatigue and failure, resulting in leaks.

Age and Wear: Like all mechanical systems, HVAC equipment deteriorates over time with regular use. Aging components, including seals, gaskets, and tubing, may degrade and develop leaks as they approach the end of their service life.

Chemical Exposure: Exposure to chemicals, solvents, or contaminants can degrade rubber seals and gaskets, causing them to deteriorate and leak refrigerant.

Improper Maintenance: Inadequate maintenance, such as neglecting regular inspections, servicing, and leak detection, can allow small leaks to go undetected and worsen over time. Detecting and repairing refrigeration leaks promptly is essential to prevent energy waste, system inefficiency, and environmental harm. Regular maintenance, including leak inspections and proper repair procedures, can help mitigate the risk of refrigerant leaks in HVAC systems.

> **Repairing refrigerant leaks in HVAC systems involves several steps and depends on the location and severity of the leak. Here's a general guide on how to repair refrigerant leaks:**

Locate the Leak: Use a refrigerant leak detector, ultrasonic leak detector, or bubble solution to locate the source of the leak. Common areas where leaks occur include joints, connections, valves, coils, and tubing.

Assess the Severity: Determine the size and severity of the leak. Small leaks may only require sealing or patching, while larger leaks may necessitate component replacement.

Isolate the System: Shut off power to the HVAC system and isolate the section of the system containing the leak. This may involve closing valves, disconnecting lines, or isolating components.

Repair options:

Sealants: For small leaks in accessible areas, such as joints or connections, sealants like epoxy or refrigerant leak sealant may be used to patch the leak.

Brazing or Soldering: For larger leaks or damaged tubing, brazing or soldering may be necessary to repair the leak. This involves heating the damaged area and applying a brazing rod or solder to create a new seal.

Replacement: In cases of severe corrosion, physical damage, or extensive leaks, it may be necessary to replace the damaged components, such as tubing, coils, valves, or fittings.

Clean the Area: Thoroughly clean the area around the leak to remove dirt, debris, and contaminants that could interfere with the repair process.

Sealants: Follow the manufacturer's instructions for applying sealants, ensuring proper adhesion and coverage of the leak area.

Brazing or Soldering: Use a torch and appropriate brazing or soldering equipment to repair the leak. Exercise caution to avoid overheating nearby components and causing further damage.

Replacement: Disconnect and remove the damaged components, then install new components according to manufacturer specifications. Ensure proper alignment, brazing, or soldering of connections.

Pressure Test: After making the repair, pressurize the system with nitrogen or dry air and perform a pressure test to verify the integrity of the repair. Check for any additional leaks before recharging the system with refrigerant.

Recharge the System: Once the repair is complete and verified, recharge the HVAC system with the appropriate refrigerant to the manufacturer's specifications.

Test and Monitor: Turn on the HVAC system and test it to ensure proper operation. Monitor the system for any signs of leaks or issues in the repaired area.

Preventive Measures: Implement preventive measures, such as regular maintenance, leak detection, and inspections, to prevent future refrigerant leaks and ensure the continued efficiency and performance of the HVAC system.

SUPERHEAT AND SUBCOOLING

Superheat and subcooling are essential measurements used to diagnose the efficiency and performance of HVAC systems, particularly air conditioning and refrigeration systems. Here's how you can check superheat and subcooling:

Superheat: Superheat is the temperature increase of a vapor above its saturation temperature at a particular pressure. It indicates the amount of sensible heat added to the refrigerant vapor after it has fully evaporated in the evaporator coil.

To check superheat: Attach a thermometer or thermocouple to the suction line (the line returning refrigerant vapor to the compressor) near the outlet of the evaporator coil. Measure the temperature of the refrigerant vapor at this point. Use a pressure-temperature chart or a digital manifold gauge set to determine the saturation temperature of the refrigerant at the same pressure. Subtract the saturation temperature from the actual temperature to calculate the superheat. The result is the amount of sensible heat in the refrigerant vapor. Superheat should typically fall within a specified range, depending on factors such as the type of refrigerant, system design, and operating conditions.

Subcooling: Subcooling is the temperature decrease of a liquid refrigerant below its saturation temperature at a particular pressure. It indicates the amount of sensible heat removed from the refrigerant liquid after it has fully condensed in the condenser coil.

To check subcooling: Attach a thermometer or thermocouple to the liquid line (the line supplying liquid refrigerant to the expansion device) near the outlet of the condenser coil. Measure the temperature of the refrigerant liquid at this point. Use a pressure-temperature chart or a digital manifold gauge set to determine the saturation temperature of the refrigerant at the same pressure. Subtract the actual temperature from the saturation temperature to calculate the subcooling. The result is the amount of sensible heat removed from the refrigerant liquid. Like superheat, subcooling should typically fall within a specified range, depending on system design and operating conditions.

Interpreting Readings: Superheat and subcooling values can vary depending on factors such as outdoor temperature, indoor temperature, humidity, airflow, and system load. Compare measured superheat and subcooling values to manufacturer specifications or industry standards for the specific refrigerant and system type. Abnormal superheat or subcooling readings may indicate issues such as improper refrigerant charge, airflow problems, clogged filters, refrigerant leaks, or component malfunctions.

HOW TO USE A MULTIMETER

Using a multimeter involves several steps to measure various electrical parameters accurately. Here's a general guide on how to use a multimeter:

Select the Proper Function and Range: Turn the dial on the multimeter to select the desired function. Common functions include voltage (V), current (A), resistance (Ω), and continuity (usually denoted by a sound wave symbol). Choose the appropriate range for the measurement you're making. Start with the highest range and adjust downward as needed for better accuracy.

Set Up the Test Leads: For voltage and resistance measurements, connect the red test lead to the VΩ terminal (positive) and the black test lead to the common (COM) terminal (negative). For current measurements, switch the red test lead to the current (A) terminal on the multimeter.

Turn On the Multimeter: If your multimeter has a power switch, turn it on. Some multimeters may turn on automatically when the leads are connected to a circuit.

Perform the measurement:

For voltage measurements: Touch the red test lead to the positive (+) terminal of the component or circuit being measured and the black test lead to the negative (-) terminal. Read the voltage value displayed on the multimeter screen.

For current measurements: Break the circuit and connect the multimeter in series with the load (in line with the current flow). Be sure to select the appropriate current range on the multimeter to avoid damaging the meter.

For resistance measurements: Disconnect the power to the circuit being measured. Touch the test leads to the two points across which you want to measure resistance. Read the resistance value displayed on the multimeter screen.

For continuity measurements: Touch the two test leads together. If the circuit is continuous (i.e., there is a complete path for current flow), the multimeter will emit a beep or display a low resistance reading.

Interpret the Measurement: Ensure that the measured value falls within the expected range for the parameter being measured. Consider the units (volts, amps, ohms) and scale of the measurement to interpret the result correctly.

Turn Off the Multimeter: When you've finished using the multimeter, turn it off to conserve battery life.

Safety Precautions: Always follow proper safety precautions when working with electrical circuits. Avoid touching exposed conductors or components while the circuit is energized. Use appropriate personal protective equipment (PPE) as necessary.

> **By following these steps and practicing proper safety procedures, you can use a multimeter effectively to measure various electrical parameters in circuits and components.**

HOW TO USE HVAC GAUGES

Using HVAC gauges, also known as manifold gauges or refrigeration gauges, is essential for diagnosing and troubleshooting HVAC systems. Here's how to use HVAC gauges: Select the Proper Gauge Set: Ensure you have the correct gauge set for the type of refrigerant used in the HVAC system you're working on (e.g., R-410A, R22, etc.). Different refrigerants require gauge sets with specific hoses, fittings, and pressure scales to ensure accurate readings.

Prepare the HVAC System: Ensure the HVAC system is turned off before connecting the gauge set. If necessary, recover or evacuate any refrigerant from the system before performing maintenance or repairs.

Connect the Gauge Set: Attach the blue hose to the low-pressure port on the HVAC system. The low-pressure side is typically the larger suction line that connects to the evaporator coil. Connect the red hose to the high-pressure port on the HVAC system. The high-pressure side is usually the smaller liquid line that connects to the condenser coil. Ensure that the yellow hose, which is connected to the center port of the manifold gauge, is open to the atmosphere.

Open Valves and Purge Hoses: Open both the low-pressure (blue) and high-pressure (red) valves on the manifold gauge set. Slowly crack open the valves on the HVAC system to allow refrigerant to flow into the gauge set. This purges any air from the hoses. Close the valves on the HVAC system once refrigerant begins to flow steadily from the hoses.

Take Pressure Readings: Check the pressure readings on both the low-pressure and high-pressure gauges. The readings will vary depending on factors such as ambient temperature, system load, and refrigerant type. Compare the pressure readings to the manufacturer's specifications or industry standards to ensure the system is operating within the recommended range.

Interpret Pressure Readings: Low-pressure (suction) readings indicate the condition of the evaporator coil and the amount of refrigerant returning to the compressor. High-pressure (discharge) readings provide information about the condenser coil and the pressure of refrigerant leaving the compressor.

Perform Additional Tests: Use the gauge set to perform additional tests, such as checking for system leaks, measuring subcooling and superheat, and troubleshooting system performance issues.

Disconnect the Gauge Set: Once you've completed your measurements and diagnostics, close the valves on the manifold gauge set to stop the flow of refrigerant. Disconnect the hoses from the HVAC system, ensuring that any remaining refrigerant is captured or recovered according to environmental regulations.

Store the Gauge Set Properly: After use, store the gauge set in a clean, dry location to protect it from damage and contamination. Ensure that the hoses are properly coiled and secured to prevent kinks or leaks.

By following these steps and practicing proper safety procedures, you can use HVAC gauges effectively to diagnose and troubleshoot HVAC systems. Remember to always follow manufacturer instructions and industry best practices when working with refrigerants and HVAC equipment.

HOW TO USE HVAC DIGITAL GAUGES

Using HVAC digital gauges, also known as digital manifold gauges, provides accurate measurements of refrigerant pressures, temperatures, and other system parameters. Here's how to use HVAC digital gauges:

Select the Proper Gauge Set: Ensure you have the correct digital manifold gauge set for the type of refrigerant used in the HVAC system you're working on (e.g., R-410A, R-22, etc.).

> **Different refrigerants require specific gauge sets with appropriate fittings, sensors, and pressure scales.**

Power On the Digital Gauge: Turn on the digital manifold gauge by pressing the power button or following the manufacturer's instructions. Some digital gauges may require calibration or initialization before use. Follow the manufacturer's guidelines for proper setup.

Connect the Hoses: Attach the hoses included with the digital gauge set to the low-pressure (suction) and high-pressure (discharge) ports on the HVAC system. Ensure that the hose fittings match the ports on the system and that they are securely attached to prevent leaks.

Open Valves and Purge Hoses: Open the valves on the digital manifold gauge set to allow refrigerant to flow into the system. Slowly open the valves on the HVAC system to purge any air from the hoses and ensure a clean refrigerant flow. Close the valves on the HVAC system once refrigerant begins to flow steadily from the hoses.

Take Pressure and Temperature Readings: Use the digital gauge display to monitor pressure readings on both the low-pressure (suction) and high-pressure (discharge) sides of the HVAC system. Some digital gauges also provide temperature measurements. Use the temperature sensors included with the gauge set to measure temperatures at various points in the system, such as at the evaporator and condenser coils.

Interpret Readings and Diagnose Issues: Compare pressure and temperature readings to manufacturer specifications or industry standards to determine if the system is operating within the recommended range. Use the digital gauge's diagnostic features, such as pressure differential measurements, superheat, and subcooling calculations, to diagnose system performance issues and troubleshoot problems.

Perform Additional Tests: Utilize the digital gauge's additional features, such as leak detection, vacuum measurement, and refrigerant charging capabilities, to perform comprehensive system tests and maintenance tasks.

Store the Digital Gauge Properly: After use, turn off the digital gauge and disconnect the hoses from the HVAC system. Store the digital gauge and accessories in a clean, dry location to protect them from damage and contamination.

By following these steps and referring to the manufacturer's instructions, you can effectively use HVAC digital gauges to measure and analyze system parameters, diagnose issues, and perform maintenance tasks on HVAC systems.

Refrigerator Repairs

TEMPERATURE FLUCTUATIONS

Temperature fluctuations in a refrigerator can be frustrating and may lead to food spoilage if not addressed promptly. Several common causes can contribute to temperature fluctuations:

Dirty Condenser Coils: Over time, dust, dirt, and debris can accumulate on the condenser coils located either at the back or underneath the refrigerator. When these coils are dirty, they cannot dissipate heat efficiently, leading to temperature fluctuations. Regular cleaning of the condenser coils is essential to maintain optimal cooling performance.

Faulty Evaporator Fan Motor: The evaporator fan circulates cold air from the evaporator coils throughout the refrigerator and freezer compartments. If the fan motor is faulty or fails to operate correctly, it can lead to uneven cooling and temperature fluctuations. A malfunctioning evaporator fan motor may need to be replaced to restore proper cooling.

Damaged Door Gaskets: The door gaskets create a seal around the refrigerator and freezer doors to prevent warm air from entering and cold air from escaping. If the door gaskets are damaged, worn, or dirty, they may not seal properly, allowing warm air to infiltrate the refrigerator and cause temperature fluctuations. Inspect the door gaskets regularly and replace them if necessary.

Faulty Temperature Control Thermostat: The temperature control thermostat regulates the refrigerator's cooling system by turning the compressor on and off as needed to maintain the desired temperature. If the thermostat is faulty or out of calibration, it may not maintain consistent temperatures, leading to fluctuations. A defective thermostat may require replacement to resolve the issue.

Inadequate Air Circulation: Proper airflow is essential for even cooling throughout the refrigerator compartments. Blocked air vents, obstructed evaporator coils, or overcrowded shelves can impede airflow and result in temperature variations. Ensure that air vents are not blocked and leave space around the evaporator coils for adequate airflow.

Refrigerant Leak: A refrigerant leak can disrupt the cooling process and cause temperature fluctuations in the refrigerator. Signs of a refrigerant leak may include reduced cooling capacity, frost buildup on the evaporator coils, or hissing sounds coming from the refrigerator. If a refrigerant leak is suspected, it's essential to contact a qualified technician to locate and repair the leak and recharge the refrigerant.

Faulty Defrost System: If the defrost system, including the defrost heater, defrost thermostat, or defrost timer, malfunctions, it can lead to excessive frost buildup on the evaporator coils. This can obstruct airflow and cause temperature fluctuations. Test the components of the defrost system and replace any faulty parts as needed.

Repairing temperature fluctuations in a refrigerator depends on the specific cause of the issue. Here's a step-by-step guide to troubleshooting and repairing common problems:

Clean the Condenser Coils: Turn off the refrigerator and unplug it from the power source. Locate the condenser coils, which are typically found on the back or underneath the refrigerator. Use a vacuum cleaner with a brush attachment or a condenser coil brush to remove dust, dirt, and debris from the coils. Once the coils are clean, plug in the refrigerator and monitor its temperature to see if the issue is resolved.

Check the Evaporator Fan Motor: Open the refrigerator and freezer compartments and listen for the sound of the evaporator fan running. If the fan is not running or is making unusual noises, it may be faulty and in need of replacement. Remove any obstruction that may be blocking the fan blades and preventing them from turning. If the fan motor is defective, replace it with a new one following the manufacturer's instructions.

Inspect the Door Gaskets: Check the door gaskets for signs of damage, wear, or tearing. Clean the door gaskets with warm, soapy water to remove any dirt or debris that may be preventing a proper seal. If the gaskets are damaged or worn, replace them with new ones to ensure a tight seal around the refrigerator and freezer doors.

Test the Temperature Control Thermostat: Use a multimeter to test the temperature control thermostat for continuity. If the thermostat does not have continuity when set to the desired temperature, it may be faulty and need replacement.

Ensure Adequate Air Circulation: Check for blocked air vents or obstructed evaporator coils inside the refrigerator and freezer compartments. Rearrange items inside the refrigerator to ensure there is adequate space for airflow around the shelves and compartments. Remove any items that may be blocking the airflow from the vents.

Address Refrigerant Leak: If you suspect a refrigerant leak, contact a qualified appliance technician to locate and repair the leak. Refrigerant leaks require specialized equipment and training to repair safely and effectively.

Inspect and Test the Defrost System: Test the defrost heater, defrost thermostat, and defrost timer for continuity using a multimeter. Replace any defective components as needed to restore proper function of the defrost system.

Monitor Temperature: After making any repairs or adjustments, monitor the temperature inside the refrigerator and freezer compartments to ensure they are maintaining the desired temperatures consistently.

ICE MAKER MALFUNCTIONS

Ice maker malfunctions in GE refrigerators can occur due to various issues. Here are some common problems:

Clogged Water Inlet Valve: The water inlet valve supplies water to the ice maker. If it becomes clogged with debris or mineral deposits, water flow to the ice maker can be restricted or blocked entirely, resulting in slow or no ice production.

Faulty Water Inlet Valve: A malfunctioning water inlet valve can prevent water from entering the ice maker, leading to no ice production. This can occur due to electrical issues, mechanical failure, or wear and tear over time.

Frozen Water Supply Line: If the water supply line to the ice maker becomes frozen, water cannot flow properly, resulting in no ice production. This can happen if the freezer temperature is too low or if the water inlet valve is faulty.

Defective Ice Maker Assembly: The ice maker assembly itself may be defective, preventing it from properly cycling to produce ice. Common issues include broken gears, malfunctioning switches, or damaged motor components.

Clogged or Frozen Ice Maker Fill Tube: The fill tube carries water from the water inlet valve to the ice maker. If it becomes clogged with ice or debris, water flow to the ice maker can be obstructed, leading to no ice production.

Improper Water Pressure: Insufficient water pressure can prevent the water inlet valve from opening fully, resulting in reduced ice production. Conversely, excessively high water pressure can cause leaks or damage to the water inlet valve.

Faulty Ice Level Sensor: Some ice makers are equipped with a sensor that detects when the ice bin is full. If this sensor malfunctions, it may incorrectly signal that the bin is full, preventing the ice maker from producing more ice.

Electrical Issues: Problems with the electrical connections, wiring, or control board associated with the ice maker can cause malfunctions. This may include loose connections, damaged wiring, or a faulty control board.

Inadequate Freezer Temperature: If the freezer temperature is too high, it can affect the ice maker's ability to produce ice. Ensure that the freezer is set to the proper temperature (usually around 0°F/-18°C) to facilitate ice production.

Water Filter Issues: A clogged or expired water filter can restrict water flow to the ice maker, affecting ice production. Replace the water filter according to the manufacturer's recommendations to maintain proper water flow.

Repairing an ice maker malfunction in a refrigerator typically involves identifying the specific cause of the problem and addressing it accordingly. Here's a step-by-step guide to repairing common ice maker issues:

Inspect and Clean the Water Inlet Valve: Turn off the water supply to the refrigerator. Locate the water inlet valve usually located at the back of the refrigerator. Remove any debris or mineral deposits that may be blocking the valve screen using a brush or compressed air. If the valve is visibly damaged or malfunctioning, it may need to be replaced.

Check the Water Supply Line: Inspect the water supply line for any kinks, bends, or obstructions. Thaw any frozen sections of the water supply line using a hairdryer or by turning off the refrigerator and allowing it to defrost. Ensure that the water supply line is properly connected to the water inlet valve and the ice maker.

Test the Water Inlet Valve: Use a multimeter to test the continuity of the water inlet valve. If the valve does not have continuity when energized, it is faulty and needs to be replaced.

Inspect and Clean the Fill Tube: Locate the fill tube that carries water from the water inlet valve to the ice maker. Check for any clogs or obstructions in the fill tube that may be preventing water flow. Use a flexible wire or pipe cleaner to remove any debris or ice buildup from the fill tube.

Test the Ice Maker Components: Check the ice maker assembly for any visible signs of damage or malfunction, such as broken gears or a faulty motor. Test the motor, thermostat, and other components of the ice maker assembly for proper operation using a multimeter. If any components are defective, they may need to be replaced to restore the proper ice maker function.

Ensure Proper Freezer Temperature: Check the temperature of the freezer compartment and adjust it to the recommended setting if necessary (usually around 0°F/-18°C). Ensure that the freezer vents are not blocked and that airflow is unobstructed.

Replace the Water Filter: If the refrigerator is equipped with a water filter, replace it according to the manufacturer's recommendations to ensure proper water flow to the ice maker.

Test the Ice Maker: After making any necessary repairs or adjustments, turn the ice maker back on and monitor its operation. Wait for the ice maker to cycle and fill with water, then check for proper ice production.

Monitor Performance: Monitor the ice maker's performance over the next few days to ensure that the issue has been resolved and that ice production is consistent.

FROST BUILDUP

Excessive frost buildup in a refrigerator, particularly in the freezer compartment, can lead to several issues.

Poor Cooling Performance: When frost accumulates excessively, it can insulate the evaporator coils, reducing their ability to cool the air properly. This can result in the refrigerator and freezer compartments not maintaining their set temperatures, leading to food spoilage.

Increased Energy Consumption: As the refrigerator works harder to maintain the desired temperature due to the insulating effect of the frost, it consumes more energy, leading to higher electricity bills.

Difficulty in Closing Doors: Thick frost buildup can interfere with the proper sealing of refrigerator and freezer doors. This compromises the efficiency of the appliance and can lead to further frost accumulation.

Ice Maker Malfunctions: Excessive frost can hinder the operation of the ice maker, causing it to produce less ice or malfunction altogether.

Water Leakage: When frost melts, it can turn into water that leaks onto the floor, potentially causing damage to the surrounding area.

Common causes of excessive frost buildup include:

Defective Defrost System: If the defrost heater, defrost timer, or defrost thermostat malfunctions, it can prevent the system from properly melting frost off the evaporator coils.

Door Seal Issues: Damaged or worn door seals allow warm, humid air from outside to enter the refrigerator, leading to condensation and frost buildup.

Temperature Settings: If the refrigerator's temperature settings are too low, it can cause excessive frost accumulation.

Airflow Obstructions: Blocked air vents or shelves can disrupt airflow inside the refrigerator, leading to uneven cooling and potential frost buildup.

> **Repairing excessive frost buildup in a refrigerator involves identifying the underlying cause and taking appropriate steps to address it. Here's a general guide:**

Identify the Cause: Start by determining why frost is accumulating excessively. Check for common issues such as a malfunctioning defrost system, damaged door seals, blocked air vents, or improper temperature settings.

Inspect the Defrost System: If the defrost system is suspected to be the cause, examine components such as the defrost heater, defrost timer, and defrost thermostat for any signs of malfunction. You may need to use a multimeter to test for continuity in these components.

Replace Defective Parts: If any components of the defrost system are found to be faulty, replace them as needed. This could involve installing a new defrost heater, defrost timer, or defrost thermostat. Refer to the refrigerator's manual or consult with a professional technician for guidance on part replacement.

Check Door Seals: Inspect the door seals for any signs of damage, wear, or accumulation of debris. Replace any damaged seals to ensure a tight seal between the doors and the refrigerator cabinet.

Clear Air Vents: Ensure that air vents inside the refrigerator and freezer compartments are not blocked by food items or other obstructions. This allows for proper airflow and prevents uneven cooling.

Adjust Temperature Settings: Verify that the refrigerator and freezer temperature settings are appropriate for the contents being stored. Adjust the settings if necessary to prevent frost buildup.

Clean the Interior: Remove any excess frost buildup inside the refrigerator and freezer compartments using a plastic scraper or spatula. Avoid using sharp objects that could damage the appliance.

Monitor Performance: After completing the repairs, monitor the refrigerator's performance over the next few days to ensure that the excessive frost buildup issue has been resolved. Check for proper cooling, minimal frost accumulation, and any signs of recurring problems.

WATER LEAKAGE

Water leakage in a refrigerator can be indicative of several common issues.

Clogged or Frozen Defrost Drain: If the defrost drain is clogged or frozen, water from the defrost cycle can accumulate and leak out of the refrigerator. This often manifests as water pooling in the bottom of the refrigerator or flowing onto the floor.

Faulty Water Inlet Valve: The water inlet valve supplies water to the refrigerator's ice maker and water dispenser. If it becomes defective or develops a leak, water can leak out onto the floor or into the refrigerator.

Damaged or Worn Door Seals: Worn or damaged door seals can allow warm air to enter the refrigerator, leading to condensation and water leakage. This typically occurs around the edges of the doors.

Blocked or Misaligned Defrost Drain Pan: The defrost drain pan collects water from the defrost cycle and directs it to the drain. If the drain pan is blocked or misaligned, water may overflow and leak onto the floor.

Clogged or Frozen Water Line: If the water line supplying the ice maker and water dispenser becomes clogged or frozen, it can lead to water leakage inside the refrigerator or at the back of the appliance.

Condensation: Excessive condensation inside the refrigerator can result in water pooling on shelves and in drawers. This can occur due to frequent door openings, high humidity levels, or improper temperature settings.

Cracked or Damaged Water Filter Housing: If the water filter housing is cracked or damaged, it may leak water onto the floor or into the refrigerator compartment.

Ice Maker Malfunctions: Issues with the ice maker, such as a malfunctioning ice maker assembly or damaged ice maker fill tube, can cause water to leak inside the refrigerator.

Repairing water leakage in a refrigerator typically involves identifying the source of the leak and taking appropriate corrective action. Here's a step-by-step guide to help you repair water leakage:

Locate the Source of the Leak: Begin by identifying where the water is coming from. Check inside the refrigerator, under the appliance, and around any water supply lines or components such as the ice maker and water dispenser.

Clear the Defrost Drain: If the water is pooling inside the refrigerator or leaking onto the floor, it may be due to a clogged or frozen defrost drain. Locate the drain and clear any debris or ice buildup using a small brush, pipe cleaner, or hot water. Ensure that the drain is clear and that water can flow freely through it.

Inspect the Water Inlet Valve: If the leak is coming from the back of the refrigerator or near the water supply line, check the water inlet valve for leaks or damage. Replace the water inlet valve if it's defective or if there are visible signs of leakage.

Check the Door Seals: Inspect the door seals for any cracks, tears, or gaps that could allow warm air to enter the refrigerator. Clean the seals with warm, soapy water and ensure they make a tight seal when the doors are closed. Replace any damaged door seals to prevent further leakage.

Verify the Defrost Drain Pan: Ensure that the defrost drain pan is properly positioned beneath the evaporator coils and is not cracked or damaged. If the drain pan is misaligned or damaged, water may overflow and leak onto the floor. Adjust or replace the drain pan as needed.

Thaw Frozen Water Lines: If the water supply line to the ice maker or water dispenser is frozen, carefully thaw it using a hairdryer or warm towels. Once the line is thawed, check for any obstructions and ensure water can flow freely through the line.

Monitor for Condensation: If water leakage is due to condensation inside the refrigerator, reduce humidity levels by adjusting temperature settings or using a dehumidifier. Wipe up any excess moisture inside the refrigerator to prevent water accumulation.

Test the Refrigerator: After making repairs, monitor the refrigerator for any signs of leakage over the next few days. Check all connections, seals, and components to ensure they are functioning properly and that the leak has been resolved.

UNUSUAL NOISES

Unusual noises in a refrigerator can be indicative of various issues. Here are some common causes:

Compressor Noise: The compressor is responsible for circulating refrigerant through the system to cool the refrigerator. If it's making loud or unusual noises, such as buzzing, humming, or rattling, it could indicate a malfunctioning compressor motor, worn bearings, or refrigerant issues.

Condenser Fan Issues: The condenser fan helps dissipate heat from the refrigerator's condenser coils. If it's malfunctioning or obstructed, it may produce unusual noises like buzzing, grinding, or clicking.

Evaporator Fan Problems: The evaporator fan circulates cold air inside the refrigerator and freezer compartments. If it's damaged or obstructed, it may make loud noises such as buzzing, squealing, or clicking.

Defrost Timer Clicking: The defrost timer controls the refrigerator's defrost cycle. Clicking sounds from the defrost timer are normal during operation. However, if the clicking becomes frequent or irregular, it may indicate a problem with the timer.

Water Inlet Valve Noise: If your refrigerator has a water dispenser or ice maker, the water inlet valve may produce humming or buzzing noises when it's filling with water. Excessive noise could indicate a malfunctioning valve.

Ice Maker Noises: If your refrigerator has an ice maker, it may produce various sounds during the ice-making process, such as humming, buzzing, or clunking. However, if the noises are excessive or unusual, it could indicate a problem with the ice maker assembly or water supply line.

Damaged or Worn Components: Loose or damaged components within the refrigerator, such as shelves, drawers, or door bins, may produce rattling, clanking, or squeaking noises during operation.

Frost or Ice Buildup: Excessive frost or ice buildup inside the refrigerator or freezer compartments can cause cracking or popping sounds as the ice expands and contracts during defrost cycles.

Repairing unusual noises in a refrigerator depends on the specific cause of the noise. Here's a general guide to help you address common issues:

Identify the Source of the Noise: Try to locate where the noise is coming from. Is it the compressor, condenser fan, evaporator fan, ice maker, or another component? This will help you narrow down the possible causes.

Inspect and Clean: Once you've identified the source of the noise, inspect the component for any visible signs of damage, obstruction, or wear. Clean the area around the component to remove any debris or dust that could be causing the noise.

Tighten Loose Components: If the noise is coming from loose components such as shelves, drawers, or door bins, tighten any screws or fasteners to secure them in place. This may help reduce rattling or clanking noises.

Check the Condenser and Evaporator Fans: If the noise is coming from the condenser or evaporator fan, inspect the fan blades for damage or obstructions. Clear any debris that may be obstructing the fan's movement. If the fan motor is making unusual noises, it may need to be lubricated or replaced.

Inspect the Compressor: If the noise is coming from the compressor, check for loose mounting bolts or damaged rubber mounts. Tighten any loose bolts and replace damaged mounts if necessary. If the compressor itself is making unusual noises, it may indicate a more serious issue that requires professional repair.

Examine the Ice Maker: If the noise is coming from the ice maker, check the water supply line and inlet valve for leaks or obstructions. Ensure that the ice maker is properly installed and aligned. If the noise persists, the ice maker assembly may need to be repaired or replaced.

Adjust Temperature Settings: Sometimes, adjusting the temperature settings can help reduce noise caused by the compressor or other components. Experiment with different temperature settings to see if the noise improves.

FAULTY COMPRESSOR

A faulty compressor in a refrigerator can lead to several issues, including:

Inadequate Cooling: The compressor is responsible for circulating refrigerant through the system to cool the refrigerator. If the compressor is faulty or not functioning properly, it may fail to adequately cool the refrigerator, resulting in warmer-than-desired temperatures inside the appliance.

No Cooling at All: In some cases, a completely failed compressor may result in no cooling whatsoever. This can lead to rapid food spoilage and loss of perishable items.

Clicking or Buzzing Noises: A faulty compressor may produce unusual noises such as clicking, buzzing, or rattling. These noises may occur intermittently or persistently and could indicate mechanical issues within the compressor motor or electrical problems.

Frequent Cycling: If the compressor is cycling on and off more frequently than usual, it may be a sign of compressor malfunction. This can result in temperature fluctuations inside the refrigerator and freezer compartments.

Overheating: A malfunctioning compressor may overheat during operation, leading to potential damage to the compressor motor or other components. Overheating can be caused by electrical issues, refrigerant leaks, or mechanical failures within the compressor.

Tripped Circuit Breakers or Fuses: If the compressor draws excessive current or experiences electrical faults, it may cause circuit breakers to trip or fuses to blow. This can result in a complete loss of power to the refrigerator.

Refrigerant Leaks: A faulty compressor may develop leaks in the refrigerant system, leading to a loss of cooling capacity and potential environmental hazards. Refrigerant leaks can be identified by hissing noises, oil stains around the compressor, or frost buildup on refrigerant lines.

High Energy Consumption: A malfunctioning compressor may require more energy to operate, leading to higher electricity bills. This is often due to inefficient operation or constant cycling of the compressor.

Repairing a faulty compressor in a refrigerator is a complex task that often requires specialized knowledge and tools. However, here are some general steps involved in repairing or replacing a faulty compressor:

Diagnosis: Before attempting any repairs, accurately diagnose the problem to ensure that the compressor is indeed faulty. This may involve checking for common symptoms such as inadequate cooling, unusual noises, or tripped circuit breakers. Use a multimeter to test the compressor motor windings for continuity and check for refrigerant leaks.

Safety Precautions: Always prioritize safety when working with electrical appliances. Unplug the refrigerator from the power source and allow it to discharge for several minutes before attempting any repairs. Wear appropriate safety gear, such as gloves and safety goggles, to protect yourself from electrical hazards and refrigerant exposure.

Tools and Parts: Gather the necessary tools and replacement parts for the repair. This may include a multimeter, screwdrivers, wrenches, a refrigerant recovery system, replacement compressor, and any other materials needed for the job.

Disconnect Power and Refrigerant: Disconnect the power supply to the refrigerator and discharge the refrigerant from the system using a refrigerant recovery system. Follow proper procedures to safely recover and dispose of the refrigerant according to local regulations.

Remove the Old Compressor: Remove any components obstructing access to the compressor, such as panels, brackets, or tubing. Disconnect the electrical connections and refrigerant lines from the compressor. Carefully remove the compressor from its mounting position.

Install the New Compressor: Install the new compressor in place of the old one, ensuring proper alignment and connection of electrical wires and refrigerant lines. Use new gaskets and fittings as needed to prevent leaks.

Recharge Refrigerant: If the refrigerant system was discharged during the repair, recharge it with the appropriate type and amount of refrigerant. Follow manufacturer guidelines and best practices for refrigerant charging to ensure proper performance and efficiency.

Test and Verify: After completing the repair, test the refrigerator to ensure that the compressor is functioning correctly and that the cooling system is operating as intended. Monitor temperature levels and check for any signs of leaks or abnormal operation.

Reassemble and Secure: Reassemble any panels, brackets, or components that were removed during the repair. Ensure that all connections are secure and properly tightened to prevent leaks or electrical hazards.

CONDENSER COILS

Common issues with condenser coils on a refrigerator include:

Dirt and Debris Buildup: Over time, dust, pet hair, and other debris can accumulate on the condenser coils, reducing their ability to dissipate heat efficiently. This can lead to poor cooling performance and higher energy consumption.

Restricted Airflow: Blocked or obstructed airflow around the condenser coils can impede their ability to release heat. This can occur if the refrigerator is placed too close to the wall or if there are objects blocking the airflow around the coils.

Condenser Fan Malfunction: The condenser fan helps to circulate air over the condenser coils to aid in heat dissipation. If the fan motor fails or the fan blades become damaged, it can result in inadequate cooling and higher temperatures inside the refrigerator.

Refrigerant Leak: A refrigerant leak in the refrigeration system can cause the condenser coils to operate inefficiently. Signs of a refrigerant leak include frost or ice buildup on the coils, hissing noises, or a decrease in cooling performance.

Damage to Coils: Physical damage to the condenser coils, such as bending or puncturing, can impair their ability to transfer heat effectively. This can occur due to accidental impact or improper handling during maintenance or cleaning.

Corrosion: Corrosion on the condenser coils can occur over time, especially in humid environments or if the coils are exposed to moisture. Corrosion can reduce the efficiency of the coils and lead to cooling problems.

Improper Maintenance: Neglecting routine maintenance, such as cleaning the condenser coils and inspecting the condenser fan, can contribute to issues with the coils over time.

Repairing issues with condenser coils on a refrigerator involves a combination of cleaning, troubleshooting, and, if necessary, replacing components. Here's a step-by-step guide on how to repair common condenser coil issues:

Safety Precautions: Before you start any repair work, unplug the refrigerator from the power outlet to avoid the risk of electric shock. Additionally, wear appropriate safety gear, such as gloves and safety glasses, to protect yourself during the repair process.

Access the Condenser Coils: Locate the condenser coils on your refrigerator. Depending on the model, they may be located on the back of the refrigerator or underneath behind a grille. Use a screwdriver to remove any panels or covers obstructing access to the coils.

Clean the Condenser Coils: Use a soft brush or a vacuum cleaner with a brush attachment to gently remove dust, dirt, and debris from the condenser coils. Work carefully to avoid bending or damaging the coils. You can also use a coil cleaning brush or a specialized coil cleaning solution for more thorough cleaning.

Check for Obstructions: Inspect the area around the condenser coils for any obstructions that may restrict airflow, such as debris, pet hair, or objects blocking the ventilation openings. Remove any obstructions to ensure proper airflow to the coils.

Inspect the Condenser Fan: Check the condenser fan located near the condenser coils to ensure that it's operating properly. Spin the fan blades by hand to check for smooth rotation and listen for any unusual noises. If the fan is not functioning correctly, it may need to be cleaned or replaced.

Address Refrigerant Leaks: If you suspect a refrigerant leak, it's essential to contact a qualified technician to inspect and repair the leak. Refrigerant handling requires specialized equipment and training, so it's not something you should attempt to repair yourself.

Reassemble the Refrigerator: Once you've completed the repairs, reassemble any panels or covers that were removed to access the condenser coils. Make sure all connections are secure and properly tightened.

Plug In and Test: Plug the refrigerator back into the power outlet and turn it on. Monitor the refrigerator's cooling performance and temperature levels to ensure that the repairs have resolved the issue.

Regular Maintenance: To prevent future issues with the condenser coils, make it a habit to clean them regularly, at least once or twice a year. This will help maintain your refrigerator's optimal performance and efficiency.

DOOR SEAL

Common issues with door seals on a refrigerator include:

Wear and Tear: Over time, the door seals (also known as gaskets) can become worn, torn, or cracked due to frequent use, age, or exposure to heat and moisture. This compromises their ability to create a tight seal, leading to air leaks.

Debris Accumulation: Dust, food particles, and debris can accumulate along the edges of the door seals, preventing them from making proper contact with the refrigerator cabinet. This allows warm air to enter the refrigerator, leading to temperature fluctuations and energy inefficiency.

Loose or Misaligned Seals: If the door seals are loose or misaligned, they may not make proper contact with the refrigerator cabinet when the doors are closed. This can result in gaps or leaks that allow cold air to escape and warm air to enter the refrigerator.

Hardening or Compression: Door seals can harden or lose their elasticity over time, especially if they are exposed to extreme temperatures or cleaning agents. This can prevent them from conforming to the contours of the refrigerator cabinet and creating an airtight seal.

Mold and Mildew Growth: Moisture trapped in the door seals can create an ideal environment for mold and mildew growth. This not only compromises the integrity of the seals but also poses a health risk if not addressed promptly.

Freezer Door Seal Issues: In refrigerators with separate freezer compartments, the door seals on the freezer door may develop issues such as tears, gaps, or frost buildup. This can lead to frost accumulation inside the freezer and reduced cooling efficiency.

Repairing door seals on a refrigerator involves several steps to address common issues and restore the seals' effectiveness. Here's a guide on how to repair door seals:

Inspect the Seals: Start by visually inspecting the door seals for any signs of wear, tear, gaps, or damage. Look closely along the edges and corners of the seals to identify any issues.

Clean the Seals: Use a mild detergent or a solution of water and vinegar to clean the door seals thoroughly. Gently scrub the seals with a soft sponge or cloth to remove dirt, debris, and any buildup. Rinse the seals with clean water and wipe them dry with a towel.

Check for Proper Alignment: Ensure that the door seals are properly aligned and making full contact with the refrigerator cabinet when the doors are closed. Close the doors and inspect the seals to see if there are any gaps or areas where the seals are not touching the cabinet.

Adjust Alignment if Necessary: If the door seals are not properly aligned, you may need to adjust the hinges or latches to ensure a snug fit. Use a screwdriver or wrench to loosen the screws on the hinges, adjust the alignment as needed, and tighten the screws to secure the hinges in place.

Use a Heat Gun or Hairdryer: If the door seals are misshapen or have lost their elasticity, you can use a heat gun or a hairdryer to soften the seals and reshape them. Apply heat evenly along the length of the seals until they become pliable, then gently mold them into the desired shape. Allow the seals to cool and set in the new shape before closing the doors.

Apply Silicone Sealant: If there are small tears or gaps in the door seals, you can use silicone sealant to patch them. Apply a thin layer of silicone sealant along the affected areas and smooth it out with your finger or a spatula. Allow the sealant to dry completely before closing the doors.

Replace Worn Seals: If the door seals are severely worn, torn, or damaged beyond repair, they will need to be replaced. Contact the manufacturer or a reputable appliance parts supplier to obtain replacement seals that are compatible with your refrigerator model. Follow the manufacturer's instructions or consult a professional technician for guidance on replacing the seals.

Regular Maintenance: Make it a habit to inspect and clean the door seals regularly to prevent dirt, debris, and mold from accumulating. This will help prolong the lifespan of the seals and ensure that they remain effective in sealing the refrigerator doors.

ELECTRICAL PROBLEMS

Common electrical issues with refrigerators can include:

Power Supply Problems: Issues with the power supply, such as a tripped circuit breaker or a blown fuse, can cause the refrigerator to lose power. This may be due to electrical overloads, faulty wiring, or problems with the outlet.

Faulty Power Cord: A damaged or frayed power cord can lead to intermittent power loss or electrical shorts. Inspect the power cord for any signs of damage, such as exposed wires or cuts, and replace it if necessary.

Defective Start Relay or Overload Protector: The start relay and overload protector are electrical components that help start and protect the compressor. If these components fail, the compressor may not start properly, resulting in cooling issues or complete failure of the refrigerator.

Malfunctioning Compressor: A faulty compressor can cause various electrical problems, such as overheating, excessive power consumption, or failure to cool properly. This may be due to issues with the compressor motor, electrical connections, or internal components.

Damaged Wiring or Connections: Wiring damage or loose connections within the refrigerator's electrical system can lead to electrical shorts, power loss, or malfunctioning components. Inspect the wiring and connections for any signs of wear, corrosion, or damage, and repair or replace them as needed.

Faulty Control Board: The control board regulates various electrical functions of the refrigerator, such as temperature control, defrost cycles, and fan operation. If the control board malfunctions, it can cause erratic behavior or complete failure of the refrigerator.

Defective Defrost Heater or Thermostat: Problems with the defrost heater or thermostat can lead to frost buildup in the freezer compartment, reduced cooling efficiency, or overheating of the refrigerator. This may be due to electrical failures or sensor malfunctions.

Ice Maker Issues: Electrical problems with the ice maker, such as malfunctioning solenoid valves, faulty water inlet valves, or defective control modules, can cause issues with ice production, water leaks, or electrical shorts.

Interior Lighting Problems: If the interior lights of the refrigerator are not working, it may be due to electrical issues such as blown bulbs, faulty light sockets, or damaged wiring.

Electronic Display Malfunctions: Refrigerators with electronic displays or control panels may experience electrical issues such as display glitches, error codes, or unresponsive buttons.

> Repairing electrical issues with a refrigerator can be complex and potentially hazardous due to the involvement of electrical components. Here's a general guide on how to repair common electrical problems with a refrigerator:

Safety First: Before you start any repairs, unplug the refrigerator from the power outlet to avoid the risk of electric shock. Additionally, wear appropriate safety gear, such as insulated gloves and safety goggles, to protect yourself from electrical hazards.

Diagnose the Problem: Begin by identifying the specific electrical issue with the refrigerator. This may involve inspecting components, testing electrical connections, and using a multimeter to check for continuity or voltage.

Check Power Supply: Verify that the refrigerator is receiving power by checking the circuit breaker or fuse box for any tripped breakers or blown fuses. Reset or replace any faulty breakers or fuses as needed.

Inspect Power Cord: Examine the power cord for any signs of damage, such as fraying, cuts, or exposed wires. If the power cord is damaged, replace it with a new one to ensure safe operation.

Examine Electrical Components: Inspect electrical components such as the start relay, overload protector, control board, defrost heater, thermostat, and compressor for any signs of damage or malfunction. Look for burnt or melted parts, loose connections, or corrosion.

Test Components: Use a multimeter to test the electrical continuity and voltage of various components, such as the compressor, start relay, defrost heater, and thermostat. Replace any faulty components that are not functioning properly.

Replace Defective Parts: If you identify any defective electrical components, replace them with new ones that are compatible with your refrigerator model. Follow manufacturer guidelines and instructions for proper installation.

Check Wiring and Connections: Inspect the wiring harness and electrical connections within the refrigerator for any signs of damage, corrosion, or loose connections. Repair or replace any damaged wiring or connectors as needed.

Perform System Checks: Once you've made the necessary repairs and replacements, perform system checks to ensure that the refrigerator is functioning properly. Monitor temperature levels, check for proper operation of the compressor and fans, and verify that the electrical components are working as intended.

Reassemble and Test: Reassemble any panels or covers that were removed during the repair process and plug the refrigerator back into the power outlet. Test the refrigerator to ensure that the electrical issues have been resolved and that it is operating safely and efficiently.

Regular Maintenance: To prevent future electrical problems, perform regular maintenance on your refrigerator, including cleaning, inspecting electrical components, and addressing any issues promptly.

CONDENSATION

Condensation in a refrigerator can be indicative of several common issues:

Humidity Levels: Fluctuating humidity levels inside the refrigerator can lead to condensation forming on the walls, shelves, and food containers. This is particularly common during humid weather or if the refrigerator door is frequently opened and closed.

Door Seal Problems: Worn or damaged door seals can allow warm, moist air from the surrounding environment to enter the refrigerator, leading to condensation buildup. Inspect the door seals for any signs of wear, tears, or gaps and replace them if necessary.

Improper Temperature Settings: If the refrigerator temperature is set too low, it can cause excess moisture to accumulate inside the appliance, leading to condensation. Ensure that the temperature settings are adjusted to the manufacturer's recommended levels.

Blocked Air Vents: Blocked air vents inside the refrigerator can disrupt airflow and cause temperature imbalances, leading to condensation formation. Make sure that the air vents are not obstructed by food items or containers.

Warm Food Items: Placing warm or hot food items directly into the refrigerator can increase the internal temperature and humidity, promoting condensation. Allow hot foods to cool to room temperature before placing them in the refrigerator.

Excessive Frost or Ice Buildup: Excessive frost or ice buildup inside the refrigerator or freezer compartments can lead to condensation when it melts. Check for any signs of frost or ice accumulation and defrost the appliance if necessary.

Poor Refrigerator Ventilation: Inadequate ventilation around the refrigerator can trap warm air and moisture, contributing to condensation. Ensure that there is sufficient space between the refrigerator and surrounding walls or cabinets for proper airflow.

Faulty Door Gaskets or Hinges: If the refrigerator doors do not close properly due to faulty door gaskets or hinges, it can allow warm air to enter the appliance, leading to condensation. Inspect the door gaskets and hinges for any damage and repair or replace them as needed.

Refrigerator Overloading: Overloading the refrigerator with too many items can restrict airflow and cause temperature fluctuations, leading to condensation. Avoid overcrowding the refrigerator and maintain adequate space between food items for proper air circulation.

Repairing condensation issues in a refrigerator typically involves addressing the underlying causes and taking appropriate corrective measures. Here's a step-by-step guide on how to repair condensation problems:

Inspect Door Seals: Start by examining the door seals for any signs of wear, tear, gaps, or damage. Ensure that the seals are making full contact with the refrigerator cabinet when the doors are closed. Replace the door seals if they are damaged or not sealing properly.

Clean Door Seals: Use a mild detergent or a solution of water and vinegar to clean the door seals thoroughly. Remove any dirt, debris, or food particles that may be preventing the seals from forming a tight seal. Rinse the seals with clean water and dry them thoroughly with a towel.

Check Temperature Settings: Verify that the refrigerator temperature is set to the manufacturer's recommended levels. Adjust the temperature settings if necessary to prevent excessive cooling and condensation buildup inside the appliance.

Inspect Air Vents: Check for any obstructions or blockages in the air vents inside the refrigerator. Ensure that the vents are not obstructed by food items or containers, as this can disrupt airflow and lead to condensation formation.

Defrost the Refrigerator: If there is excessive frost or ice buildup inside the refrigerator or freezer compartments, manually defrost the appliance to remove the accumulated ice. Follow the manufacturer's instructions for defrosting the refrigerator safely.

Ensure Proper Ventilation: Maintain adequate ventilation around the refrigerator to allow for proper airflow and prevent trapped heat and moisture. Ensure that there is sufficient space between the refrigerator and surrounding walls or cabinets.

Inspect for Refrigerant Leaks: Check for any signs of refrigerant leaks, such as hissing noises, oil stains, or frost buildup on refrigerant lines. If you suspect a refrigerant leak, contact a professional technician to inspect and repair the leak.

Monitor Humidity Levels: Keep track of humidity levels inside the refrigerator and adjust as needed. If humidity levels are too high, consider using a dehumidifier or moisture absorber inside the appliance to help reduce condensation.

Regular Maintenance: Perform regular maintenance on your refrigerator, including cleaning, inspecting door seals, and checking for signs of wear or damage. Address any issues promptly to prevent condensation problems from worsening.

Electric Stove

BURNER MALFUNCTION

Common issues with burner malfunction on an electric stove can include:

Burner Not Heating Up: The burner fails to heat up at all, which can be caused by a variety of factors such as a faulty heating element, burner socket, control switch, or electrical wiring problem.

Uneven Heating: The burner heats unevenly, with some areas getting hotter than others. This can result from a damaged or warped heating element, poor contact with the cookware, or electrical issues.

Burner Overheating: The burner heats up excessively, even when set to a low temperature. This can be caused by a malfunctioning control switch or thermostat, which fails to regulate the temperature properly.

Burner Not Maintaining Temperature: The burner heats up initially but fails to maintain the desired temperature, causing fluctuations in heat output. This can be due to issues with the control switch, thermostat, or heating element.

Burner Sparking or Arcing: There may be sparking or arcing coming from the burner element, indicating a problem with the electrical connections or the heating element itself. This poses a safety hazard and should be addressed immediately.

Burner Element Damage: The burner element may be visibly damaged, with signs such as cracks, breaks, or discoloration. This can affect its ability to heat properly and may require replacement.

Control Switch Malfunction: The control switch that regulates the burner's temperature may be faulty, causing erratic behavior or failure to turn the burner on or off.

Burner Socket Issues: The burner socket, which connects the burner element to the stove's electrical system, may be damaged or corroded, leading to poor contact and unreliable heating.

Electrical Wiring Problems: Issues with the electrical wiring supplying power to the burner can cause malfunctions such as intermittent heating, power surges, or complete failure to heat.

Cookware Compatibility: Some issues with burner performance may be related to the type or quality of cookware being used. Poorly conductive or warped cookware can affect heat distribution and cause uneven heating.

> **Repairing burner malfunction on an electric stove involves several steps to diagnose and address the underlying issues. Here's a guide on how to repair common burner problems:**

Safety Precautions: Before you begin any repairs, ensure that the stove is turned off and unplugged from the power source to prevent the risk of electric shock. Additionally, wear protective gloves to avoid burns.

Visual Inspection: Inspect the burner element for any visible signs of damage, such as cracks, breaks, or discoloration. Also check the burner socket and wiring for signs of corrosion or loose connections.

Clean Burner and Socket: If there is any dirt, grease, or residue on the burner element or socket, clean it thoroughly using a damp cloth or a mild detergent solution. Make sure the burner and socket are completely dry before reassembly.

Check Control Switch and Thermostat: Test the control switch and thermostat for proper functionality. Use a multimeter to check for continuity and resistance across the terminals of the switch and thermostat. Replace them if they are faulty.

Inspect Electrical Connections: Verify that the electrical connections supplying power to the burner are secure and undamaged. Tighten any loose connections and replace any damaged wiring or connectors.

Test Burner Operation: Turn on the stove and test the burner to see if it heats up properly and maintains the desired temperature. If the burner still malfunctions, proceed to the next step.

Replace Burner Element: If the burner element is damaged or defective, it will need to be replaced. Refer to the manufacturer's instructions for your stove model to ensure proper replacement. Disconnect the old burner element and install the new one in its place.

Replace Burner Socket: If the burner socket is corroded or damaged, it should be replaced. Disconnect the old socket and install the new one, making sure to secure the electrical connections properly.

Test the Stove: After completing the repairs, test the stove again to ensure that the burner is functioning correctly and that there are no further issues.

Regular Maintenance: To prevent future burner malfunctions, clean the burner elements and sockets regularly and inspect them for signs of wear or damage. Additionally, avoid using cookware that is too large or small for the burners, as this can affect heating efficiency.

UNEVEN HEATING

Uneven heating in an electric stove can lead to cooking inconsistencies and frustration. Here are some common issues that can cause uneven heating:

Warping or Damage to Burner Element: Over time, the burner element on an electric stove can warp or become damaged, leading to uneven heating. This can result in hot spots or areas that do not heat evenly.

Poor Contact with Cookware: If the cookware is not making good contact with the burner element, it can lead to uneven heating. This can happen if the cookware is warped, too small for the burner, or if the bottom is not flat.

Incorrect Burner Size: Using a burner that is too large or too small for the size of the cookware can cause uneven heating. The burner should match the size of the bottom of the pot or pan to ensure even heat distribution.

Dirty or Damaged Burner Socket: A dirty or damaged burner socket can prevent the burner element from making proper contact with the stove, leading to uneven heating. Cleaning the socket or replacing it if damaged can help resolve this issue.

Faulty Control Switch: A faulty control switch can cause uneven heating by not regulating the temperature of the burner properly. If the switch is defective, it may not send the correct amount of power to the burner, leading to hot spots or areas that do not heat evenly.

Electrical Issues: Problems with the electrical wiring or connections supplying power to the burner can cause uneven heating. Loose connections or damaged wiring can lead to fluctuations in power, resulting in uneven heating.

Cookware Material: Certain types of cookware, such as thin or lightweight pans, may not distribute heat evenly. Using high-quality, heavy-bottomed cookware can help improve heat distribution and reduce uneven heating.

> Repairing uneven heating in an electric stove involves identifying the underlying cause and taking appropriate corrective measures. Here's a step-by-step guide on how to address this issue:

Inspect Burner Element: Check the burner element for any signs of damage, warping, or discoloration. If the element is visibly damaged, it may need to be replaced. Ensure that the burner element is seated correctly in the burner socket.

Check Cookware: Ensure that the cookware being used is suitable for the burner size and material of the cooktop. Avoid using warped or lightweight cookware, as they can contribute to uneven heating. Use heavy-bottomed pots and pans for better heat distribution.

Clean Burner Socket: Turn off the stove and allow it to cool completely. Carefully remove the burner element and clean the socket beneath it using a soft brush or cloth. Make sure there is no debris or residue that could interfere with proper contact between the element and the socket.

Test Control Switch: Turn on the affected burner and adjust the temperature using the control switch. If the burner does not respond or heats unevenly, the control switch may be faulty and need to be replaced. Test the other burners to see if they are experiencing similar issues.

Inspect Electrical Connections: Turn off the power to the stove at the circuit breaker or fuse box. Carefully remove the back panel of the stove to access the electrical connections. Check for any loose or damaged wires and tighten or replace them as needed.

Test Burner Operation: Turn the power back on and test the burner again to see if the issue has been resolved. Place a level on the cookware to ensure that it is sitting flat on the burner and making good contact.

Replace Burner Element or Control Switch: If the burner element or control switch is found to be faulty during testing, it will need to be replaced. Refer to the stove's manual or contact the manufacturer for replacement parts and instructions on how to install them.

BROKEN CONTROL KNOBS OR SWITCHES

Common issues with broken control knobs or switches on an electric stove can include:

Knobs Not Turning: The control knobs may become stuck or difficult to turn, preventing you from adjusting the temperature or turning the stove on and off.

Knobs Breaking: The knobs themselves may break or become damaged due to wear and tear, impact, or excessive force when turning.

Switches Not Functioning: The control switches behind the knobs may malfunction, causing them to fail to regulate the temperature or turn the stove elements on and off.

Loose or Wobbly Knobs: The knobs may become loose or wobbly, making it difficult to accurately set the desired temperature or function.

Knob Shaft Damage: The shafts or stems that connect the knobs to the control switches may become damaged or stripped, preventing them from properly engaging with the switches.

Electrical Shorts: If the control switches or wiring behind the knobs become damaged, it can lead to electrical shorts, causing the stove to malfunction or pose a safety hazard.

To repair broken control knobs or switches on an electric stove, consider the following steps:

Turn Off Power: Before starting any repairs, ensure that the power to the stove is turned off at the circuit breaker or fuse box to prevent the risk of electric shock.

Remove Knobs: Carefully remove the broken control knobs from the stove by pulling them straight off the shafts or stems.

Inspect Knobs and Shafts: Examine the knobs and shafts for any signs of damage, such as cracks, breaks, or stripping. If the knobs are broken, they will need to be replaced. If the shafts are damaged, they may need to be repaired or replaced as well.

Check Control Switches: Behind each knob, locate the corresponding control switch. Test the switches for proper functionality by turning them with pliers or a screwdriver. If a switch is faulty, it may need to be replaced.

Replace Knobs and Shafts: Install new knobs onto the shafts or stems, ensuring that they are securely attached. If the shafts are damaged, replace them with new ones that are compatible with the stove model.

Secure Knobs: Once the knobs and shafts are in place, ensure that the knobs are securely attached and turn smoothly without sticking or wobbling.

Restore Power and Test: Turn the power back on at the circuit breaker or fuse box and test the repaired control knobs and switches to ensure that they are functioning properly.

FAULTY OVEN TEMPERATURE

Common issues with a faulty oven temperature in an electric stove include:

Inaccurate Temperature: The oven may not reach the set temperature, or it may heat up beyond the set temperature, leading to uneven cooking or burned food.

Temperature Fluctuations: The oven temperature may fluctuate during operation, resulting in inconsistent cooking results.

Slow Preheating: The oven may take longer than usual to preheat to the desired temperature, prolonging cooking times.

Failure to Reach High Temperatures: The oven may fail to reach high temperatures required for certain cooking tasks, such as baking or broiling.

Uneven Heating: Different areas of the oven may heat unevenly, leading to uneven cooking or browning of food.

Thermostat Calibration Issues: The oven's thermostat may be out of calibration, causing it to inaccurately regulate the temperature.

Faulty Temperature Sensor: The temperature sensor inside the oven may be malfunctioning or damaged, preventing it from accurately detecting the oven temperature.

Oven Door Seal Problems: A damaged or worn oven door seal can allow heat to escape from the oven, affecting its ability to maintain the desired temperature.

Control Panel Malfunctions: Issues with the control panel, such as faulty buttons, switches, or electronic components, can affect the oven's temperature regulation.

To repair a faulty oven temperature in an electric stove, consider the following steps:

Check Oven Thermometer: Use an oven thermometer to measure the actual temperature inside the oven and compare it to the set temperature. This can help determine if the oven is heating accurately.

Calibrate Thermostat: Refer to the oven's manual for instructions on how to calibrate the thermostat. This may involve adjusting the temperature offset to match the actual oven temperature.

Inspect Temperature Sensor: Inspect the temperature sensor inside the oven for any signs of damage or corrosion. If the sensor is faulty, it may need to be replaced.

Test Heating Elements: Test the oven heating elements for continuity using a multimeter. If a heating element is faulty, it may need to be replaced.

Check Oven Door Seal: Inspect the oven door seal for any signs of damage, wear, or gaps. Replace the seal if necessary to ensure proper heat retention.

Clean Oven Interior: Remove any built-up food debris or grease from the oven interior, as this can affect heat distribution and temperature regulation.

Reset Oven Controls: Reset the oven controls by turning off the power to the stove at the circuit breaker or fuse box for a few minutes, then turning it back on.

SELF-CLEANING

Self-cleaning features on electric stoves can encounter various issues, including:

Incomplete Cleaning: The self-cleaning cycle may not thoroughly clean the oven interior, leaving behind residue or baked-on stains.

Smoke or Odors: During the self-cleaning cycle, smoke or unpleasant odors may be emitted from the oven, which can be caused by burning food debris or grease.

Excessive Heat: The oven may become excessively hot during the self-cleaning cycle, causing surrounding cabinets or countertops to become damaged or discolored.

Door Lock Problems: The oven door may fail to lock or unlock properly during the self-cleaning cycle, preventing the cycle from starting or completing.

Control Panel Malfunctions: Issues with the control panel or electronic components can prevent the self-cleaning cycle from initiating or completing successfully.

Timer or Programming Errors: Errors in the programming or timer settings can disrupt the self-cleaning cycle, leading to incomplete cleaning or premature cycle termination.

Oven Damage: The intense heat generated during the self-cleaning cycle can cause damage to oven components, such as heating elements, temperature sensors, or door seals.

Safety Concerns: Self-cleaning cycles can pose safety risks, such as fire hazards or carbon monoxide emissions, if not used properly or if there are issues with the oven's ventilation system.

To repair common issues with self-cleaning on an electric stove, consider the following steps:

Check Oven Interior: Before starting the self-cleaning cycle, remove any large food debris or spills from the oven interior to prevent excessive smoking or odors.

Verify Oven Door Lock: Ensure that the oven door locks properly before starting the self-cleaning cycle. Clean the door latch mechanism if it is dirty or obstructed.

Reset Control Panel: If the control panel is unresponsive or displaying error codes, try resetting it by turning off the power to the stove at the circuit breaker or fuse box for a few minutes, then turning it back on.

Clean Oven Ventilation System: Check the oven ventilation system, including vents and filters, for any obstructions or buildup of debris. Clean or replace filters if necessary to ensure proper airflow during the self-cleaning cycle.

Monitor Cycle Progress: During the self-cleaning cycle, monitor the oven for any signs of excessive heat, smoke, or odors. If necessary, stop the cycle and allow the oven to cool before investigating further.

ELECTRICAL WIRING PROBLEMS

Common issues with electrical wiring on electric stoves can include:

Loose Connections: Over time, electrical connections within the stove can become loose due to vibrations or thermal expansion and contraction. Loose connections can lead to intermittent power supply or electrical shorts.

Damaged Wiring: Wiring within the stove can become damaged due to wear and tear, rodents, or accidental damage. Damaged wiring can result in electrical malfunctions, such as power loss or overheating.

Corrosion: Corrosion of electrical terminals and connectors can occur over time, especially in humid environments or if exposed to moisture. Corrosion can interfere with electrical conductivity and lead to poor performance or failure of electrical components.

Burnt Wires: Excessive heat or electrical overloads can cause wires to become burnt or melted. Burnt wires can lead to power loss, electrical shorts, or even fire hazards if not addressed promptly.

Poor Insulation: Insulation on electrical wires can degrade over time, exposing bare wires and increasing the risk of electrical shocks or short circuits.

Improper Installation: Incorrect installation of electrical wiring during stove assembly or repair can result in wiring issues, such as misconnected or improperly secured wires.

Overloaded Circuits: Connecting multiple appliances to the same electrical circuit can overload the circuit and lead to electrical wiring issues, including overheating and potential fire hazards.

Rodent Damage: Rodents can chew through electrical wiring in the stove, causing electrical malfunctions or safety hazards.

To repair electrical wiring issues on electric stoves, consider the following steps:

Visual Inspection: Inspect the electrical wiring within the stove for any signs of damage, loose connections, or corrosion. Look for burnt or melted wires and ensure that all connections are secure.

Tighten Connections: If you identify any loose electrical connections, tighten them using a screwdriver or appropriate tools. Ensure that connections are tight but not over-tightened to avoid damaging components.

Repair or Replace Damaged Wiring: If you find any damaged wiring, such as burnt or melted wires, repair or replace the affected wires as needed. Use proper electrical wiring and connectors rated for stove use.

Address Corrosion: Clean any corrosion from electrical terminals and connectors using a wire brush or an appropriate cleaning solution. Apply a corrosion inhibitor to prevent future corrosion.

Ensure Proper Insulation: Replace any damaged or degraded insulation on electrical wires to prevent electrical shocks or short circuits. Use heat-resistant and flame-retardant insulation materials suitable for stove use.

Avoid Overloading Circuits: Avoid connecting multiple high-power appliances to the same electrical circuit to prevent overloading. Use dedicated circuits for high-power appliances like electric stoves.

INDICATOR LIGHT MALFUNCTIONS

Common issues with indicator light malfunctions on an electric stove can include:

Failure to Illuminate: Indicator lights may fail to illuminate when the associated burner or element is turned on, making it difficult to determine if the stove is operating.

Continuous Illumination: Indicator lights may remain lit even when the associated burner or element is turned off, potentially causing confusion or indicating a malfunction.

Flickering Lights: Indicator lights may flicker intermittently, indicating a poor electrical connection or faulty component.

Inconsistent Lighting: The brightness or color of indicator lights may vary between burners or elements, making it challenging to gauge their status accurately.

Delayed Response: Indicator lights may have a delayed response, taking longer than usual to illuminate or turn off when the corresponding burner or element is activated or deactivated.

Dim or Faint Illumination: Indicator lights may appear dim or faint, making them difficult to see, especially in bright lighting conditions.

Burnt-Out Bulbs: The indicator light bulbs may burn out over time, resulting in no illumination when the associated burner or element is activated.

To repair indicator light malfunctions on an electric stove, consider the following steps:

Inspect Indicator Light Bulbs: Check the indicator light bulbs for signs of damage or burnout. If a bulb is burnt out, replace it with a new bulb of the appropriate type and wattage.

Clean Indicator Light Covers: Clean the indicator light covers to remove any dirt, grease, or residue that may be obstructing the light. Use a soft cloth or mild detergent solution to clean the covers thoroughly.

Check Electrical Connections: Inspect the electrical connections associated with the indicator lights for any loose or corroded connections. Tighten loose connections and clean corrosion if present.

Test Indicator Light Operation: Test the indicator lights to ensure they illuminate properly when the corresponding burner or element is activated. If any lights fail to illuminate or exhibit abnormal behavior, further troubleshooting may be required.

Replace Indicator Light Switches: If indicator lights continue to malfunction after checking bulbs and connections, the associated switches may be faulty and need to be replaced. Consult the stove's manual or a professional technician for assistance with switch replacement.

BROKEN LATCH OR HINGES

Common issues with broken latches or hinges on an electric stove's oven door can include:

Difficulty Closing or Opening: A broken latch or hinge can result in difficulty closing or opening the oven door smoothly.

Misalignment: The oven door may become misaligned if the hinges are damaged or broken, causing it to not close properly or create gaps between the door and the oven frame.

Door Won't Stay Closed: A broken latch or hinge may prevent the oven door from staying closed during operation, leading to heat loss and inefficient cooking.

Door Won't Open: In some cases, a broken latch or hinge can prevent the oven door from opening, trapping food inside the oven.

Door Falls Open: If a hinge is broken, the oven door may fall open unexpectedly, posing a safety hazard and potentially causing injury.

Broken Latch Mechanism: The latch mechanism that secures the oven door closed may become broken or damaged, preventing it from engaging properly.

To repair common issues with broken latches or hinges on an electric stove's oven door, consider the following steps:

Inspect Latches and Hinges: Examine the latches and hinges on the oven door for any signs of damage, such as cracks, breaks, or misalignment.

Clean and Lubricate: Clean the hinges and latch mechanism to remove any dirt, debris, or rust that may be causing the issue. Apply a small amount of lubricant to the hinges to help them move smoothly.

Tighten Screws: Check the screws and bolts securing the hinges to the oven frame and the door. Tighten any loose screws or bolts to ensure a secure connection.

Replace Broken Parts: If a latch or hinge is broken or damaged beyond repair, it will need to be replaced. Contact the manufacturer or a professional appliance repair technician to obtain replacement parts and install them correctly.

Adjust Hinges: If the oven door is misaligned, you may be able to adjust the hinges to realign it. Refer to the oven's manual or consult a professional technician for guidance on how to adjust the hinges properly.

Test Door Operation: After repairing or replacing the broken latch or hinge, test the oven door to ensure that it opens and closes smoothly and securely.

THE STOVE WON'T TURN ON

Common issues when an electric stove won't turn on can stem from various factors, including:

Power Supply Problems: Check if there's power reaching the stove. Ensure the stove is plugged into a working outlet and check the circuit breaker or fuse box to ensure the stove's circuit hasn't tripped or blown a fuse.

Faulty Power Cord: Inspect the power cord for any visible damage, such as fraying or cuts. If the power cord is damaged, it may need to be replaced.

Faulty Control Panel: Issues with the stove's control panel, such as malfunctioning buttons, switches, or electronic components, can prevent the stove from turning on. Check for any error codes or indicators on the control panel display.

Broken Control Switch: If the stove uses knobs or switches to control the burners or oven, a broken or malfunctioning control switch can prevent the stove from turning on. Test the switches for continuity using a multimeter.

Defective Heating Element: If a specific burner or the oven isn't turning on, it could be due to a defective heating element. Test the heating elements for continuity using a multimeter to determine if they're functioning properly.

Safety Features Engaged: Some electric stoves have safety features, such as child locks or overheating protection, that can prevent the stove from turning on. Check the stove's manual for instructions on how to reset or disable these features.

Faulty Thermal Fuse: A blown thermal fuse can interrupt the power supply to the stove and prevent it from turning on. Test the thermal fuse for continuity using a multimeter and replace it if necessary.

Internal Wiring Issues: Internal wiring problems, such as loose connections or damaged wires, can prevent the stove from receiving power. Inspect the internal wiring and connections for any signs of damage or corrosion.

Electronic Control Board Failure: The electronic control board controls the operation of the stove, and a failure in this component can prevent the stove from turning on. Test the control board for proper functionality and replace it if necessary.

To repair an electric stove that won't turn on, follow these steps:

Check Power Supply: Ensure the stove is plugged into a functioning power outlet. If it's plugged in and there's still no power, check the circuit breaker or fuse box to see if the stove's circuit has tripped or blown a fuse. Reset or replace as needed

Inspect Power Cord: Examine the power cord for any visible damage such as fraying or cuts. If damaged, replace the power cord with a new one compatible with your stove model.

Test Control Panel: If your stove has a digital control panel, check for any error codes or indicators on the display. If buttons or switches are used, ensure they're not stuck or malfunctioning. Clean the control panel and buttons if dirty.

Test Control Switches: For stoves with knobs or switches, test each control switch for continuity using a multimeter. Replace any switches that don't show continuity when turned on.

Inspect Heating Elements: If specific burners or the oven aren't turning on, test the heating elements for continuity using a multimeter. Replace any heating elements that don't show continuity.

Check Safety Features: Refer to the stove's manual to understand any safety features like child locks or overheating protection. Reset or disable these features as needed.

Test Thermal Fuse: Test the thermal fuse for continuity using a multimeter. If it's blown, replace it with a new one compatible with your stove model.

Inspect Internal Wiring: Turn off power to the stove and carefully inspect internal wiring and connections for damage or corrosion. Tighten any loose connections and replace any damaged wires.

Test Electronic Control Board: If previous steps don't resolve the issue, test the electronic control board for proper functionality using a multimeter. Replace the control board if it is defective.

COOKTOP GLASS DAMAGE

Common issues of cooktop glass damage on an electric stove can include:

Cracks: Cracks can occur due to sudden temperature changes, impact from heavy objects, or manufacturing defects.

Scratches: Scratches may result from abrasive cleaning materials, sliding cookware with rough bottoms, or contact with metal utensils.

Chips: Chips can occur if heavy objects are dropped onto the cooktop or if the edges are struck with force.

Staining: Stubborn stains can accumulate over time from spills or food residues that are not cleaned promptly.

Discoloration: Prolonged exposure to high heat or contact with certain substances can cause discoloration of the cooktop glass.

Pitting: Pitting refers to small indentations or pits in the glass surface, which may occur due to abrasive materials or harsh cleaning chemicals.

To prevent cooktop glass damage:

Use Proper Cookware: Use cookware with smooth, flat bottoms that are compatible with electric stoves to minimize the risk of scratches and cracks.

Avoid Sliding Cookware: Lift cookware instead of sliding it across the cooktop surface to prevent scratches.

Use Gentle Cleaning Methods: Clean the cooktop regularly with mild soap and water or specialized cooktop cleaners using soft cloths or sponges. Avoid abrasive cleaners or scrubbing pads that can scratch the glass.

Protect the Surface: Use protective mats or trivets to prevent direct contact between cookware and the cooktop surface.

Monitor Temperature: Avoid sudden temperature changes, such as placing cold or frozen items directly onto a hot cooktop, which can cause thermal shock and lead to cracks.

> **If cooktop glass damage occurs, consider the following steps for repair or replacement:**

Assess the Damage: Evaluate the extent of the damage, including the size, location, and severity of cracks, scratches, or chips.

Temporary Solutions: For minor damage, such as small scratches or chips, consider using temporary repair kits or solutions to minimize the appearance of imperfections.

Professional Repair: For significant damage or safety concerns, contact a professional appliance technician or the stove manufacturer for repair or replacement options.

Replacement: If the damage is extensive or affects the functionality of the cooktop, replacement of the glass may be necessary. Consult the stove's manual or manufacturer for information on replacement parts and procedures.

Gas Stoves

IGNITION PROBLEMS

Common ignition problems with gas stoves can include:

Weak or No Spark: The igniter may produce a weak spark or fail to spark at all when attempting to light the burner.

Delayed Ignition: The burner may take longer than usual to ignite after turning the knob, resulting in delayed ignition.

Clicking Sound Without Ignition: The igniter may continuously emit a clicking sound without successfully igniting the burner flame.

Intermittent Ignition: The igniter may work sporadically, causing the burner flame to light inconsistently or not at all.

Pilot Light Outage: For stoves equipped with pilot lights, the pilot light may go out, preventing proper burner ignition.

Dirty or Damaged Igniter: Build-up of dirt, grease, or debris on the igniter electrode can hinder its ability to produce a spark. Additionally, physical damage to the igniter electrode can impair its functionality.

Misaligned Igniter: The igniter electrode may become misaligned relative to the burner, preventing proper spark formation.

Faulty Ignition Switch: Problems with the ignition switch or control module can prevent the igniter from receiving power or initiating the spark.

Gas Supply Issues: Insufficient gas pressure or flow to the burner can hinder ignition, leading to weak or inconsistent flames.

Electrical Problems: Issues with the electrical wiring, connections, or components within the ignition system can disrupt spark generation and ignition.

To troubleshoot ignition problems with gas stoves:

Ensure Gas Supply: Verify that the gas supply valve is open and that there are no disruptions to the gas supply to the stove.

Clean Igniter and Burner: Clean the igniter electrode and burner components thoroughly to remove any dirt, grease, or debris that may be obstructing ignition.

Check Ignition Switch: Test the ignition switch or control module to ensure it is sending power to the igniter when the burner knob is turned.

Inspect Spark: Visually inspect the igniter electrode for signs of damage or misalignment. Adjust or replace the igniter as needed.

Test Pilot Light: For stoves with pilot lights, ensure the pilot light is lit and burning steadily. Clean pilot orifices and thermocouples as needed.

Verify Electrical Connections: Check all electrical connections and wiring within the ignition system for tightness and integrity. Repair or replace any damaged components.

UNEVEN HEATING

Common issues of uneven heating in a gas stove can include:

Burner Ports Blocked: Blockage of burner ports due to food residue, grease, or debris can disrupt the flow of gas and cause uneven heating.

Burner Clogged or Dirty: A buildup of dirt, grease, or food particles on the burner itself can obstruct the flame and result in uneven heating.

Uneven Gas Distribution: Inadequate gas pressure or improper gas distribution to the burner can lead to uneven flames and heating.

Burner Misalignment: Misalignment of the burner relative to the burner ports or cooking surface can cause uneven heat distribution.

Incorrect Flame Adjustment: Improper adjustment of the burner flame, such as having too low or too high of a flame, can result in uneven heating.

Burner Component Damage: Damage to burner components, such as the burner head, burner cap, or igniter, can affect flame quality and cause uneven heating.

Gas Supply Issues: Insufficient gas pressure or flow to the burner can lead to weak or inconsistent flames and uneven heating.

Thermocouple Problems: Malfunctioning or mispositioned thermocouples can interfere with proper burner ignition and flame stability, resulting in uneven heating.

To address uneven heating in a gas stove:

Clean Burner Ports: Use a soft brush or toothpick to clean burner ports and remove any food residue, grease, or debris that may be blocking them.

Clean Burner Components: Remove burner components, such as burner heads and caps, and clean them thoroughly to remove any dirt or grease buildup.

Check Gas Supply: Verify that the gas supply valve is fully open and that there are no obstructions or leaks in the gas line.

Adjust Flame: Adjust the burner flame using the burner control knob to achieve a steady, blue flame with minimal yellowing or flickering.

Inspect Burner Alignment: Ensure that the burner is properly aligned with the burner ports and cooking surface. Adjust the burner if necessary to improve heat distribution.

Test Thermocouples: Check the position and condition of thermocouples to ensure they are properly positioned and functioning correctly.

GAS LEAKS

Common issues with gas leaks in a gas stove can include:

Loose Connections: Loose connections between the gas supply line, regulator, or burner valves can result in gas leaks.

Damaged Gas Lines: Wear and tear, corrosion, or physical damage to gas lines can cause leaks to develop.

Faulty Burner Valves: Malfunctioning or damaged burner valves can fail to fully shut off the flow of gas, leading to leaks.

Cracked or Damaged Tubing: Cracks or damage to the flexible gas tubing connecting the stove to the gas supply can allow gas to escape.

Poor Installation: Improper installation of the gas stove, including incorrect fittings or connections, can lead to gas leaks.

Seal Degradation: Deterioration of seals, gaskets, or O-rings within the stove's components can result in gas leaks.

Pilot Light Issues: Problems with the pilot light, such as a weak flame or improper adjustment, can cause gas leaks.

External Factors: External factors such as seismic activity, shifting foundations, or accidental damage to gas lines can lead to gas leaks.

To address gas leaks in a gas stove:

Check for Odor: Natural gas is odorless, but gas companies add a distinctive odorant to help detect leaks. If you smell gas (a rotten egg odor), evacuate the area immediately.

Turn Off Gas: Shut off the gas supply to the stove at the main gas valve or gas meter.

Ventilate the Area: Open doors and windows to allow fresh air to circulate and ventilate the area.

Do Not Use Open Flames: Do not use any open flames, sparks, or electrical devices in the vicinity of a suspected gas leak.

Check Connections: Inspect all gas connections, fittings, and tubing for signs of damage, corrosion, or looseness. Tighten loose connections and replace damaged components as needed.

Test With Soapy Water: Mix a solution of water and dish soap and apply it to the gas connections with a brush or sponge. If bubbles form, it indicates a gas leak.

FLAME IRREGULARITIES

Common issues with flame irregularities on a gas stove can include:

Yellow Flames: Yellow flames can indicate incomplete combustion, which may be caused by insufficient air supply, clogged burner ports, or dirty burner components.

Flickering Flames: Flickering flames may result from inconsistent gas flow, air drafts around the burner, or an obstructed gas supply line.

"Lazy" Flames: "Lazy" or weak flames that fail to reach their full height can be caused by low gas pressure, a partially closed gas valve, or a blocked burner orifice.

Lifting Flames: Flames that lift off the burner surface can be caused by excessive gas pressure, improper burner adjustment, or incorrect air-to-fuel ratio.

Noisy Flames: Noisy flames, such as roaring or hissing sounds, may indicate excessive gas pressure, air entrainment, or burner misalignment.

Sooting or Blackening: Sooting or blackening of burner components or cookware can result from incomplete combustion, leading to the release of carbon particles.

Uneven Flame Distribution: Uneven flame distribution across burner ports can lead to uneven heating and cooking, caused by clogged or misaligned burner components.

To address flame irregularities on a gas stove:

Clean Burner Components: Clean burner ports, burner heads, and burner caps to remove any dirt, grease, or debris that may be obstructing the flow of gas.

Adjust Air-to-Fuel Ratio: Adjust the air shutter or Venturi tube on the burner to achieve the correct air-to-fuel ratio for optimal combustion. Consult the stove's manual for guidance on adjusting burner air mixture.

Check Gas Supply: Verify that the gas supply valve is fully open and that there are no obstructions or leaks in the gas line. Ensure proper gas pressure and flow to the burner.

Inspect Burner Alignment: Ensure that the burner is properly aligned with the burner ports and cooking surface. Adjust the burner if necessary to improve flame distribution.

Clean Gas Supply Line: Inspect the gas supply line for any blockages or obstructions that may be affecting gas flow. Clean or clear any debris as needed.

Test Pilot Light: For stoves with pilot lights, ensure the pilot light is lit and burning steadily. Clean pilot orifices and thermocouples as needed.

BURNER CLOGS

Common issues with burner clogs in a gas stove can include:

Blocked Burner Ports: Accumulation of food residue, grease, or debris can clog the small openings (ports) on the burner, obstructing the flow of gas and affecting flame quality.

Clogged Burner Heads: A buildup of dirt, grease, or food particles on the burner heads can obstruct the burner flame and result in uneven heating or flame irregularities.

Blocked Burner Orifices: The small openings (orifices) in the burner where gas is released can become clogged with debris, restricting gas flow and causing weak or uneven flames.

Dirt or Dust Accumulation: Dust, dirt, or other contaminants can accumulate inside the burner assembly over time, leading to clogs and affecting burner performance.

Insect Infestation: Insects such as spiders or ants may build nests or webs inside the burner assembly, blocking gas flow and causing burner clogs.

Corrosion: Corrosion of burner components, such as burner heads or orifice holders, can lead to the formation of rust or mineral deposits that obstruct gas flow.

To address burner clogs in a gas stove:

Turn Off Gas: Shut off the gas supply to the stove at the main gas valve or gas meter.

Remove Burner Components: Remove the burner caps, burner heads, and any other removable burner components from the stove.

Inspect for Debris: Inspect the burner ports, burner heads, and orifice holders for any signs of debris, dirt, or clogs. Use a soft brush, a toothpick, or compressed air to dislodge and remove any obstructions.

Clean Burner Components: Clean the burner components thoroughly with warm, soapy water and a soft brush or sponge to remove grease, dirt, and debris. Rinse and dry the components thoroughly before reinstalling them.

Clear Burner Orifices: Use a small, stiff wire or a specialized tool to clear any clogs or blockages from the burner orifices. Be gentle to avoid damaging the orifices.

Inspect for Insects: Check inside the burner assembly for signs of insect infestation, such as nests or webs. Use a vacuum cleaner or compressed air to remove any insects or nests.

Check Gas Supply: Verify that the gas supply valve is fully open and that there are no obstructions or leaks in the gas line. Ensure proper gas pressure and flow to the burner.

CONTROL KNOB ISSUES

Common control knob issues on a gas stove can include:

Knob Misalignment: The control knob may become misaligned with the burner valve stem, preventing it from turning the valve properly.

Knob Stuck or Difficult to Turn: The control knob may become stuck or difficult to turn due to debris, grease, or damage to the knob or valve stem.

Loose Knob: The control knob may become loose over time, causing it to wobble or feel unstable when turned.

Knob Shaft Damage: The shaft of the control knob or the valve stem it connects to may become damaged or worn, affecting the knob's ability to engage with the valve.

Knob Indicator Misalignment: The indicator markings on the control knob may become misaligned with the settings on the stove, making it difficult to determine the selected setting.

Broken Knob Shaft: The shaft of the control knob may break or snap off, preventing it from turning the burner valve.

Knob Failure to Ignite: The control knob may fail to ignite the burner when turned, indicating issues with the ignition system or gas supply.

Cracked or Damaged Knob: The control knob itself may become cracked, chipped, or otherwise damaged, affecting its functionality and appearance.

To address control knob issues on a gas stove:

Clean Knob and Valve: Use a mild detergent solution and a soft brush or cloth to clean the control knob and valve stem, removing any dirt, grease, or debris that may be causing sticking or difficulty turning.

Check for Debris: Inspect the area around the control knob and valve for any debris that may be obstructing movement. Use compressed air or a vacuum cleaner to remove debris if necessary.

Tighten Knob: If the control knob is loose, tighten the screw or retaining nut on the back of the knob to secure it to the valve stem.

Inspect Knob Shaft: Check the shaft of the control knob and the valve stem for signs of damage or wear. Replace the knob or valve stem if necessary.

Realignment: If the knob indicator is misaligned, carefully adjust the position of the knob to align it with the correct settings on the stove.

Test Ignition System: If the knob fails to ignite the burner, check the ignition system for issues such as a faulty igniter or gas supply problems. Replace or repair components as needed.

Replace Knob: If the control knob is cracked, damaged, or unable to function properly, replace it with a new knob compatible with your stove model.

OVEN TEMPERATURE INACCURACIES

Common issues of oven temperature inaccuracies in a gas stove can include:

Thermostat Calibration: The oven thermostat may be out of calibration, causing it to inaccurately regulate the oven temperature.

Temperature Sensor Malfunction: A faulty temperature sensor (also known as an oven thermocouple or temperature probe) can result in incorrect temperature readings and inconsistent heating.

Heat Distribution: Uneven heat distribution within the oven cavity can lead to temperature variations, with certain areas of the oven being hotter or cooler than others.

Insufficient Preheating: Failure to preheat the oven adequately before use can result in temperature inaccuracies, as the oven may not reach the desired temperature evenly throughout.

Door Seal Issues: Damaged or worn oven door seals (gaskets) can allow heat to escape from the oven cavity, affecting temperature stability and accuracy.

Air Ventilation: Poor air ventilation around the oven can disrupt heat circulation and lead to temperature fluctuations.

Gas Pressure Fluctuations: Variations in gas pressure can impact the burner's ability to maintain a consistent flame, affecting oven temperature control.

Ignition System Problems: Issues with the oven's ignition system, such as a faulty igniter or gas supply problems, can lead to inconsistent heating and temperature inaccuracies.

To address oven temperature inaccuracies in a gas stove:

Thermostat Calibration: Verify the oven thermostat's accuracy using an oven thermometer placed inside the oven cavity. If the thermostat is out of calibration, it may need to be adjusted or replaced.

Temperature Sensor Testing: Test the oven temperature sensor for proper functionality using a multimeter. Replace the sensor if it's faulty or inaccurate.

Heat Distribution: Arrange oven racks properly to ensure even heat distribution. Avoid overcrowding the oven and blocking airflow.

Preheating: Preheat the oven for the recommended duration to allow it to reach the desired temperature evenly throughout the cavity.

Check Door Seal: Inspect the oven door seal for any signs of damage or wear. Replace the seal if it's worn or damaged to ensure a tight seal and prevent heat loss.

Ventilation: Ensure adequate ventilation around the oven to promote proper airflow and heat circulation.

Gas Pressure Regulation: Verify that the gas pressure to the stove is consistent and within the recommended range. Contact a professional if you suspect gas pressure issues.

Ignition System Maintenance: Regularly clean and maintain the oven's ignition system, including the igniter and burner components, to ensure proper operation.

PILOT LIGHT PROBLEMS

Common pilot light problems on a gas stove can include:

Pilot Light Outage: The pilot light may go out due to drafts, air flow disruptions, or thermocouple malfunctions.

Weak Pilot Flame: A weak pilot flame can result from low gas pressure, a clogged pilot orifice, or a partially obstructed gas supply line.

Intermittent Pilot Light: The pilot light may flicker or go out intermittently due to thermocouple malfunctions, gas supply issues, or drafts.

Pilot Light Won't Stay Lit: The pilot light may fail to stay lit after ignition due to thermocouple malfunctions, improper pilot flame adjustment, or gas supply problems.

Dirty or Clogged Pilot Orifice: Accumulation of dirt, dust, or debris in the pilot orifice can obstruct gas flow and prevent proper pilot ignition.

Thermocouple Malfunction: The thermocouple, which senses the presence of the pilot flame and controls gas flow, may malfunction or fail to detect the flame, resulting in a pilot light outage.

Gas Supply Issues: Problems with the gas supply, such as low gas pressure, gas leaks, or interruptions in gas flow, can affect pilot light performance.

Pilot Light Adjustment: Incorrect pilot flame adjustment or positioning can lead to weak or unstable flames that are prone to outage.

To address pilot light problems on a gas stove:

Relight Pilot Light: Follow the manufacturer's instructions to relight the pilot light. Use a long-reach lighter or match to ignite the pilot flame.

Clean Pilot Orifice: Turn off the gas supply to the stove and carefully clean the pilot orifice using compressed air, a small wire brush, or a specialized cleaning tool to remove any dirt or debris.

Check Thermocouple: Inspect the thermocouple for proper positioning and alignment with the pilot flame. Clean the thermocouple sensor tip with a soft cloth to ensure accurate flame detection.

Adjust Pilot Flame: Adjust the pilot flame to the manufacturer's recommended height using the pilot adjustment screw or knob. A steady blue flame with a small yellow tip is ideal.

Inspect Gas Supply: Verify that the gas supply valve is fully open and that there are no obstructions or leaks in the gas line. Ensure proper gas pressure and flow to the pilot light.

Replace Thermocouple: If the pilot light still won't stay lit after relighting, the thermocouple may be faulty and require replacement.

GAS ODORS

Common issues with gas odors in gas stoves can be indicative of a gas leak, which poses a serious safety hazard. Here are some potential causes:

Leaking Gas Lines: Damage, corrosion, or loose fittings in the gas supply lines leading to the stove can result in gas leaks.

Faulty Burner Valves: Malfunctioning or damaged burner valves may fail to fully shut off the flow of gas, leading to gas leaks.

Pilot Light Issues: Problems with the pilot light, such as a weak flame or improper adjustment, can cause gas leaks.

Loose Connections: Loose connections between the gas supply line, regulator, or burner valves can result in gas leaks.

Leaking Burner Tubes: Damage or deterioration of burner tubes can lead to gas leaks.

Faulty Seals or Gaskets: Worn or damaged seals or gaskets around burner assemblies or gas valves can allow gas to escape.

Improper Installation: Incorrect installation of the gas stove, including incorrect fittings or connections, can result in gas leaks.

To address gas odors in gas stoves:

Check for Odor: If you smell gas (a rotten egg odor), evacuate the area immediately and turn off the gas supply to the stove at the main gas valve or gas meter.

Do Not Use Open Flames: Do not use any open flames, sparks, or electrical devices in the vicinity of a suspected gas leak.

Ventilate the Area: Open doors and windows to allow fresh air to circulate and ventilate the area.

Inspect Gas Lines and Connections: Inspect all gas connections, fittings, and tubing for signs of damage, corrosion, or leaks. Tighten loose connections and replace damaged components as needed.

Check Pilot Light: Ensure the pilot light is lit and burning steadily. Clean pilot orifices and thermocouples as needed.

CARBON MONOXIDE EMISSIONS

Common issues with carbon monoxide (CO) emissions in gas stoves can pose significant health risks, as CO is a colorless, odorless gas that can be deadly when inhaled in high concentrations. Here are some potential causes:

Incomplete Combustion: Inefficient or incomplete combustion of natural gas or propane in the stove's burners can produce carbon monoxide gas.

Poor Ventilation: Inadequate ventilation in the kitchen or surrounding area can allow carbon monoxide to accumulate to dangerous levels, especially in enclosed spaces.

Improper Installation: Incorrect installation of the gas stove, such as inadequate clearance around the appliance or improper venting, can lead to CO emissions.

Blocked Ventilation Systems: Blocked or obstructed ventilation systems, such as range hoods or exhaust fans, can prevent the proper expulsion of combustion gases, including carbon monoxide.

Malfunctioning Appliances: Malfunctioning or poorly maintained gas stoves can produce higher levels of carbon monoxide due to issues such as burner misalignment, clogged burner ports, or faulty ignition systems.

Poorly Adjusted Burners: Improper adjustment of burner flames can result in incomplete combustion and the production of carbon monoxide gas.

Leaking Gas Lines: Gas leaks in the supply lines leading to the stove can result in the release of natural gas or propane, which can contribute to carbon monoxide emissions.

Dishwasher

POOR CLEANING PERFORMANCE

Common issues with poor cleaning performance in a dishwasher can stem from various factors, including:

Clogged Spray Arms: If the spray arms are clogged with debris or mineral deposits, they may not distribute water effectively, resulting in poor cleaning.

Insufficient Water Pressure: Low water pressure can hinder the dishwasher's ability to clean dishes thoroughly. This may be due to a problem with the water supply or a clogged inlet valve.

Inadequate Detergent: Using too little detergent or the wrong type of detergent can lead to poor cleaning performance. Ensure you're using the appropriate detergent and the correct amount for your dishwasher.

Loading Issues: Improper loading of dishes can block water spray and prevent proper cleaning. Avoid overcrowding the dishwasher and ensure dishes are positioned to allow water to reach all surfaces.

Faulty Wash Pump: A malfunctioning wash pump can result in insufficient water circulation, leading to poor cleaning. If the wash pump is failing, dishes may not receive adequate water and detergent during the wash cycle.

Clogged Filters: Dirty or clogged filters can restrict water flow and prevent debris from being effectively removed from the dishwasher. Clean or replace filters regularly to maintain optimal performance.

Hard Water Deposits: Mineral deposits from hard water can accumulate in the dishwasher, affecting cleaning performance. Consider using a dishwasher cleaner or adding a water softener to reduce mineral buildup.

Old or Expired Detergent: Detergent that is old or past its expiration date may lose effectiveness, resulting in poor cleaning. Replace detergent regularly and store it in a cool, dry place.

Faulty Rinse Aid Dispenser: A malfunctioning rinse aid dispenser can lead to poor drying and cleaning performance. Ensure the rinse aid dispenser is filled and functioning properly.

Temperature Issues: If the water temperature is too low, detergent may not dissolve properly, leading to poor cleaning. Verify that the water heater is set to the appropriate temperature for optimal cleaning performance.

To address poor cleaning performance in a dishwasher:

Inspect and Clean Spray Arms: Remove and clean the spray arms to ensure they are free from debris and mineral deposits.

Check Water Supply: Verify that the dishwasher is receiving an adequate supply of hot water and that the water pressure is sufficient.

Use Proper Detergent: Use a high-quality dishwasher detergent and follow the manufacturer's recommendations for dosage.

Load Dishes Properly: Ensure dishes are loaded properly to allow for optimal water circulation and cleaning.

Maintain Filters: Regularly clean or replace filters to prevent clogs and ensure proper water flow.

Hard Water: Consider installing a water softener or using a dishwasher cleaner to reduce mineral buildup.

Inspect Wash Pump: If cleaning performance does not improve, consider having the wash pump inspected by a qualified technician.

Perform Regular Maintenance: Schedule regular maintenance checks to identify and address potential issues before they affect cleaning performance.

WATER LEAKAGE

Common issues with water leakage in a dishwasher can be caused by various factors, including:

Door Seal Damage: A damaged or worn door seal (gasket) can allow water to leak out during the wash cycle. Inspect the door seal for cracks, tears, or deformation and replace it if necessary.

Door Misalignment: If the dishwasher door is not properly aligned or closes improperly, it may not create a watertight seal, leading to leaks. Adjust the door hinges or latch to ensure proper alignment.

Faulty Door Latch: A malfunctioning door latch may fail to secure the dishwasher door tightly, allowing water to leak out. Inspect the door latch for damage or wear and replace it if needed.

Loose or Damaged Hose Connections: Check the connections between the dishwasher and the water supply line, drain hose, and inlet valve for leaks. Tighten loose connections or replace damaged hoses to prevent leaks.

Cracked Tub: A cracked dishwasher tub can result in water leakage during the wash cycle. Inspect the interior of the dishwasher tub for cracks or damage and replace the tub if necessary.

Worn Pump Seals: The seals around the dishwasher's pump assembly can deteriorate over time, leading to water leakage. Inspect the pump seals for signs of wear or damage and replace them if needed.

Leaking Inlet Valve: The water inlet valve, which controls the flow of water into the dishwasher, may develop leaks over time. Inspect the inlet valve for signs of leakage and replace it if necessary.

Overflowing or Blocked Drain: If the dishwasher's drain is blocked or becomes overloaded with debris, water may overflow and leak out of the dishwasher. Clean the drain filter and check for obstructions to prevent blockages.

Faulty Float Switch: The float switch detects the water level inside the dishwasher and shuts off the water supply when it reaches a certain level. A malfunctioning float switch may fail to stop the water flow, leading to overfilling and leaks.

Cracked or Damaged Tub Seal: The seal around the dishwasher tub, where it meets the door, may become cracked or damaged, allowing water to leak out. Inspect the tub seal for signs of damage and replace it if necessary.

To address water leakage in a dishwasher:

Inspect Door Seal: Check the door seal for damage and replace it if necessary.

Check Hose Connections: Tighten loose hose connections or replace damaged hoses.

Inspect Tub and Pump Seals: Inspect the dishwasher tub and pump assembly for cracks or damage and replace any worn seals.

Clean Drain: Remove any debris or obstructions from the dishwasher's drain to prevent blockages.

Replace Faulty Components: Replace any faulty components such as the door latch, inlet valve, float switch, or tub seal.

DISHES NOT DRYING

Common issues with dishes not drying properly in a dishwasher can be attributed to several factors:

Rinse Aid Level: Insufficient rinse aid in the dishwasher can result in poor drying performance. Make sure the rinse aid dispenser is filled according to the manufacturer's instructions.

Water Temperature: Inadequate water temperature during the rinse and drying cycles can lead to ineffective drying. Check that the water heater is set to the appropriate temperature (usually around 120°F or 49°C).

Heating Element Malfunction: A malfunctioning heating element may fail to generate enough heat to dry the dishes. Inspect the heating element for visible damage or signs of malfunction and replace it if necessary.

Blocked Ventilation: Blocked or obstructed vent openings inside the dishwasher can prevent moist air from escaping, leading to condensation on dishes. Ensure that the vent is clear and unobstructed.

Improper Loading: Overcrowding dishes or blocking the spray arms and vents can hinder airflow and prevent effective drying. Arrange dishes in a way that allows for proper water and air circulation.

Dish Material: Some materials, such as plastic, retain moisture more than others, leading to poor drying performance. Consider using a rinse aid specifically designed for drying plastic items.

Cooling Cycle Interruption: Opening the dishwasher door immediately after the cycle completes can interrupt the cooling cycle and trap moisture inside, leading to wet dishes. Allow the dishwasher to complete its cooling cycle before opening the door.

Faulty Vent Mechanism: A malfunctioning vent mechanism may fail to release moist air properly, resulting in poor drying performance. Check the vent for any obstructions or damage, and repair or replace it if necessary.

Old or Expired Rinse Aid: Rinse aid that is old or past its expiration date may lose effectiveness, leading to poor drying results. Replace rinse aid regularly to ensure optimal performance.

Dirty Filters: Clogged or dirty filters can restrict water flow and hinder drying performance. Clean or replace the filters according to the manufacturer's instructions.

To address dishes not drying properly in a dishwasher:

Check Rinse Aid Level: Ensure the rinse aid dispenser is filled with rinse aid to the appropriate level.

Verify Water Temperature: Check that the water heater is set to the recommended temperature for optimal drying.

Inspect Heating Element: Inspect the heating element for damage or malfunction and replace it if necessary.

Clear Ventilation: Ensure that the vent openings inside the dishwasher are clear and unobstructed.

Load Dishes Correctly: Avoid overcrowding dishes and ensure proper loading to allow for adequate airflow.

Use Proper Rinse Aid: Use a rinse aid specifically designed for drying dishes, especially plastic items.

Allow Cooling Cycle: Allow the dishwasher to complete its cooling cycle before opening the door to prevent moisture buildup.

Check Vent Mechanism: Inspect the vent mechanism for any obstructions or damage, and repair or replace it if needed.

Replace Rinse Aid: Replace rinse aid regularly to ensure effectiveness.

Clean Filters: Clean or replace filters to maintain optimal water flow and drying performance.

FOUL ODORS

Foul odors in a dishwasher are often caused by the accumulation of food particles, grease, and soap scum, creating an ideal environment for bacteria and mold growth. Common issues contributing to foul odors include:

Food Debris Buildup: Food particles trapped in the filter, spray arms, and around the door gasket can decompose and emit unpleasant odors over time.

Clogged Drain: A clogged dishwasher drain can lead to stagnant water, which can develop foul odors as it sits in the dishwasher.

Improper Loading: Overloading the dishwasher or blocking the spray arms and detergent dispenser can prevent proper water circulation and cleaning, allowing food residue to accumulate and cause odors.

Standing Water: Water pooling in the bottom of the dishwasher between cycles can become stagnant and emit foul odors.

Hard Water Deposits: Mineral buildup from hard water can accumulate in the dishwasher, contributing to odors and affecting cleaning performance.

Mold and Mildew Growth: Moisture and warmth inside the dishwasher provide an ideal environment for mold and mildew to grow, leading to foul odors.

Dirty Filter: A dirty or clogged filter can trap food particles and debris, allowing them to decompose and produce unpleasant odors.

Faulty Drain Hose: A damaged or improperly installed drain hose can lead to drainage issues, causing water to back up and emit odors.

To address foul odors in a dishwasher:

Clean Filter: Remove and clean the dishwasher filter regularly to remove food debris and prevent odors.

Clean Spray Arms: Remove any debris or clogs from the spray arms to ensure proper water circulation and cleaning.

Scrub Interior: Use a sponge or brush to scrub the interior of the dishwasher, including the door gasket and detergent dispenser, to remove any buildup of food residue and soap scum.

Check Drain: Inspect the dishwasher drain for clogs or blockages and remove any debris that may be causing drainage issues.

Use Vinegar: Run a cycle with distilled white vinegar to help break down grease and soap scum and eliminate odors. Place a cup of vinegar on the top rack of the dishwasher and run a hot water cycle.

Baking Soda: Sprinkle baking soda on the bottom of the dishwasher and run a short cycle to help neutralize odors.

Regular Maintenance: Perform regular maintenance tasks such as cleaning the dishwasher interior, checking the filter, and inspecting the drain to prevent odors from recurring.

Address Hard Water: Consider using a dishwasher cleaner specifically designed to remove hard water deposits and prevent mineral buildup.

By addressing these common issues and implementing regular maintenance practices, you can effectively eliminate foul odors and keep your dishwasher smelling fresh.

NOISY OPERATIONS

Noisy operation in a dishwasher can be disruptive and may indicate underlying issues that need attention. Common causes of noisy operation include:

Loose or Worn Parts: Loose components such as screws, bolts, or washers inside the dishwasher can rattle during operation, causing noise. Additionally, worn or damaged parts like spray arms, rollers, or bearings may produce squeaking or grinding noises.

Blocked or Clogged Spray Arms: If the spray arms are blocked by debris or mineral buildup, they may not rotate smoothly, leading to noisy operation as they hit against obstructions.

Faulty Wash Pump Motor: A malfunctioning wash pump motor can produce loud or unusual noises during operation. This may indicate a problem with the motor bearings, impeller, or other internal components.

Worn or Damaged Door Hinges: Worn or damaged door hinges can cause the dishwasher door to sag or become misaligned, resulting in noisy operation as it opens and closes.

Defective Drain Pump: A defective drain pump may produce loud or rattling noises while draining water from the dishwasher.

Hard Water Deposits: Mineral deposits from hard water can accumulate on dishwasher components, such as the spray arms or filter, causing them to vibrate or produce noise during operation.

Improper Loading: Overloading the dishwasher or improperly loading dishes can cause them to bang against each other during the wash cycle, resulting in noise.

Faulty Door Seal: A worn or damaged door seal may not provide a tight seal when the dishwasher door is closed, allowing water to leak out and causing noise during operation.

To address noisy operation in a dishwasher:

Inspect and Tighten Screws: Check for loose screws, bolts, or washers inside the dishwasher and tighten them as needed to reduce rattling or vibrating noises.

Clean Spray Arms: Remove any debris or mineral buildup from the spray arms to ensure they rotate freely and do not cause noise during operation.

Check Wash Pump Motor: Inspect the wash pump motor for signs of damage or wear and replace it if necessary to eliminate noisy operation.

Inspect Door Hinges: Check the door hinges for wear or damage and replace them if necessary to ensure smooth and quiet operation.

Replace Defective Components: Replace any worn or damaged components, such as the drain pump or door seal, that may be causing noise during operation.

Address Hard Water Deposits: Use a dishwasher cleaner specifically designed to remove hard water deposits and mineral buildup from dishwasher components.

Proper Loading: Avoid overloading the dishwasher and ensure dishes are properly arranged to prevent them from banging against each other during the wash cycle.

Regular Maintenance: Perform regular maintenance tasks such as cleaning filters, inspecting spray arms, and checking for loose parts to prevent noisy operation.

DETERGENT DISPENSER ISSUES

Common issues with detergent dispensers in dishwashers can lead to improper detergent release or complete failure to dispense detergent. Some common problems include:

Clogged Dispenser: Accumulation of detergent residue or debris in the dispenser compartment can clog the mechanism, preventing proper detergent release.

Faulty Latch Mechanism: A malfunctioning latch or spring mechanism can prevent the detergent dispenser door from opening during the wash cycle, resulting in detergent remaining unused.

Detergent Blockage: Detergent may clump together or harden, blocking the dispenser's opening and preventing the detergent from dispensing properly.

Incorrect Loading: Overfilling the dispenser with detergent or using the wrong type of detergent can lead to issues with proper dispensing. Using an excessive amount of detergent can cause the dispenser to overflow or block the dispenser door.

Damaged Seal: A damaged or worn seal around the detergent dispenser can cause leaks or prevent the door from closing properly, leading to detergent not dispensing correctly.

Electrical Malfunction: In some cases, an electrical malfunction in the dishwasher's control panel or dispenser circuitry can prevent the dispenser from opening or releasing detergent.

To address detergent dispenser issues in a dishwasher:

Clean the Dispenser: Remove any detergent residue or debris from the dispenser compartment using a soft brush or cloth. Ensure that the dispenser door moves freely and is not obstructed.

Inspect the Latch Mechanism: Check for any damage or obstruction in the latch mechanism that may be preventing the dispenser door from opening. Replace any damaged components as needed.

Clear Detergent Blockage: Remove any hardened detergent or debris blocking the dispenser's opening. Use a toothpick or soft brush to clear any clogs.

Use the Correct Detergent: Ensure that you are using the correct type and amount of dishwasher detergent recommended for your dishwasher model. Avoid overfilling the dispenser.

Replace Damaged Seal: If the seal around the detergent dispenser is damaged or worn, replace it to ensure a tight seal and prevent leaks.

Check for Electrical Issues: If other troubleshooting steps do not resolve the issue, consult the dishwasher's manual for instructions on checking for electrical malfunctions.

DRAINAGE PROBLEMS

Drainage problems in a dishwasher can result in standing water at the bottom of the appliance, improper cleaning, and even leaks. Here are common issues contributing to drainage problems:

Clogged Drain Hose: A clogged or blocked drain hose can prevent water from properly draining out of the dishwasher. This blockage can occur due to food debris, grease, or mineral buildup.

Blocked Drain Filter: The dishwasher's drain filter or trap can become clogged with food particles, preventing water from draining effectively. Regular cleaning of the filter is essential to prevent this issue.

Faulty Drain Pump: A malfunctioning drain pump may fail to remove water from the dishwasher tub. This can be due to a worn-out pump motor, a damaged impeller, or an obstruction in the pump assembly.

Improper Installation: If the dishwasher drain hose is not installed correctly or is kinked, it can impede the flow of water, leading to drainage problems.

Garbage Disposal Connection: If the dishwasher drain hose is connected to a garbage disposal unit, a clog or blockage in the disposal can affect drainage.

Air Gap Issues: An air gap, if installed, may become clogged with debris, preventing proper drainage. Cleaning or replacing the air gap component may resolve this issue.

High Loop Not Installed: If the dishwasher drain hose does not have a high loop or is not installed properly, it can lead to backflow of water from the sink into the dishwasher, causing drainage issues.

Drain Pump Impeller Damage: Damage to the drain pump impeller, caused by debris or hard water buildup, can affect the pump's ability to remove water from the dishwasher.

To address drainage problems in a dishwasher:

Clean the Drain Filter: Regularly clean the dishwasher's drain filter or trap to remove any food particles or debris that may be causing a blockage.

Check the Drain Hose: Inspect the dishwasher drain hose for kinks, bends, or obstructions. Ensure that it is installed correctly and has a high loop to prevent backflow.

Clear Blockages: Use a pipe cleaner, wire, or vinegar solution to clear any blockages in the drain hose, air gap, or garbage disposal connection.

Inspect the Drain Pump: Check the drain pump for signs of damage or obstruction. Remove any debris and ensure that the pump impeller is functioning properly.

Check Air Gap: If your dishwasher is equipped with an air gap, inspect it for blockages and clean or replace it as needed.

Ensure Proper Installation: Verify that the dishwasher drain hose is installed according to the manufacturer's instructions and is not restricted in any way.

Run a Cleaning Cycle: Occasionally run a dishwasher cleaning cycle with a dishwasher cleaner to remove buildup and debris from the interior components.

CONTROL PANEL MALFUNCTION

Control panel malfunctions in a dishwasher can disrupt the appliance's operation and may prevent you from selecting wash cycles or adjusting settings. Common issues with control panels include:

Button or Touchpad Failure: Buttons or touchpad sensors on the control panel may fail to respond when pressed or touched, preventing you from selecting options or starting the dishwasher.

Display Errors: The display panel may show error codes, flicker, or remain blank, indicating a malfunction in the control board or display module.

Inability to Start: The dishwasher may not start or respond to commands from the control panel, indicating a potential issue with the control board or power supply.

Incorrect Cycle Selection: Selecting a wash cycle or setting on the control panel may result in the dishwasher running a different cycle or failing to complete the selected cycle.

Intermittent Operation: The dishwasher may exhibit intermittent operation, where it starts and stops unexpectedly during the wash cycle or fails to complete a cycle.

Unresponsive Controls: The control panel may become unresponsive, requiring you to reset the dishwasher by turning off the power or resetting the circuit breaker.

Water Level Issues: Control panel malfunctions can lead to problems with water level detection, causing the dishwasher to overfill or underfill during the wash cycle.

Control Lock Activation: In some cases, the control panel may become locked, preventing you from making changes to settings or starting the dishwasher.

Faulty Wiring or Connections: Wiring issues or loose connections between the control panel and the dishwasher's electronic components can lead to control panel malfunctions.

Component Failure: Individual components within the control panel, such as relays, switches, or the control board itself, may fail due to wear and tear or electrical issues.

To address control panel malfunctions in a dishwasher:

Power Reset: Try resetting the dishwasher by turning off the power at the circuit breaker for a few minutes, then turning it back on to see if this resolves the issue.

Inspect Buttons or Touchpad: Check for physical damage or debris on the control panel buttons or touchpad. Clean them gently with a soft cloth and ensure they are not stuck or damaged.

Check Display Panel: Inspect the display panel for error codes or unusual behavior. Refer to the dishwasher's manual for troubleshooting steps related to specific error codes.

Test Control Board: If possible, test the dishwasher's control board or electronic control unit using a multimeter to check for continuity and proper function.

Verify Power Supply: Ensure that the dishwasher is receiving power from the electrical outlet and that the power cord is securely connected.

Inspect Wiring and Connections: Check for loose or damaged wiring connections between the control panel and other electronic components. Repair or replace any damaged wiring as needed.

Control Panel Replacement: If the control panel is damaged or unresponsive and cannot be repaired, consider replacing it with a new one that's compatible with your dishwasher model.

POOR WATER FILTRATION

Poor water filtration in a dishwasher can result in inadequate cleaning performance and the redeposition of food particles and debris onto dishes. Common issues contributing to poor water filtration include:

Clogged Filters: The dishwasher's filters or screens may become clogged with food particles, grease, or mineral deposits over time, reducing water flow and filtration efficiency.

Hard Water Deposits: Mineral buildup from hard water can accumulate on dishwasher components, including filters, spray arms, and the dishwasher tub, affecting water flow and filtration.

Faulty Wash Pump: A malfunctioning wash pump may fail to circulate water effectively through the dishwasher, reducing filtration and cleaning performance.

Inadequate Water Pressure: Low water pressure can hinder the dishwasher's ability to filter and clean dishes properly. This may be due to issues with the water supply or a clogged inlet valve.

Spray Arm Blockages: Blockages in the spray arms can restrict water flow and reduce the effectiveness of water filtration and cleaning.

Improper Loading: Overloading the dishwasher or improperly loading dishes can obstruct water circulation and filtration, leading to poor cleaning performance.

Detergent Residue: Excessive detergent residue or buildup inside the dishwasher can impair water filtration and cleaning effectiveness.

Water Softener Issues: If the dishwasher is equipped with a water softener, issues with the softener system can affect water quality and filtration.

To address poor water filtration in a dishwasher:

Clean Filters: Regularly clean the dishwasher's filters or screens to remove any accumulated debris or mineral deposits. Refer to the dishwasher's manual for instructions on filter maintenance.

Descale Components: Use a dishwasher cleaner or descaling solution to remove hard water deposits from dishwasher components, including filters, spray arms, and the dishwasher tub.

Check Wash Pump: Inspect the wash pump for signs of damage or malfunction, such as unusual noise or vibration. Replace the wash pump if necessary to restore proper water circulation.

Verify Water Pressure: Ensure that the dishwasher is receiving an adequate supply of water and that the water pressure is sufficient for optimal cleaning performance.

Clear Spray Arm Blockages: Remove any debris or mineral buildup from the spray arms to ensure unrestricted water flow and proper cleaning.

Load Dishes Correctly: Avoid overloading the dishwasher and ensure that dishes are loaded properly to allow for adequate water circulation and filtration.

Use Proper Detergent Amount: Follow the manufacturer's recommendations for detergent dosage to prevent excessive detergent residue and buildup inside the dishwasher.

Maintain Water Softener: If the dishwasher is equipped with a water softener, regularly check and maintain the softener system to ensure proper water quality and filtration.

HARD WATER DEPOSITS

Hard water deposits in a dishwasher can lead to various issues, including reduced cleaning performance, limescale buildup on dishwasher components, and the appearance of spots or film on dishes. Common issues associated with hard water deposits in a dishwasher include:

Limescale Buildup: Hard water contains high levels of minerals such as calcium and magnesium, which can accumulate as limescale deposits on dishwasher components like the heating element, spray arms, and interior surfaces.

Clogged Spray Arms: Hard water deposits can accumulate inside the spray arms, blocking the spray nozzles and reducing water flow and cleaning effectiveness.

Cloudy or Spotted Dishes: Hard water can leave behind mineral deposits on dishes, glassware, and utensils, resulting in a cloudy or spotted appearance even after washing.

Foul Odors: Limescale buildup in the dishwasher can harbor bacteria and mold, leading to foul odors emanating from the appliance.

Reduced Efficiency: Limescale deposits can insulate the heating element, reducing its efficiency and prolonging the time required to heat water for washing.

Damage to Components: Over time, hard water deposits can cause damage to dishwasher components such as the heating element, pump, and spray arms, leading to premature wear and potential malfunction.

To address hard water deposits in a dishwasher:

Use Water Softener: Consider installing a water softener system to treat hard water before it enters the dishwasher. Water softeners remove calcium and magnesium ions from the water, preventing limescale buildup.

Use a Rinse Aid: Use a rinse aid in the dishwasher to help prevent the formation of water spots and improve drying performance. Rinse aids contain ingredients that help to break down mineral deposits and prevent them from adhering to dishes.

Regular Cleaning: Clean the dishwasher interior, including the spray arms, filters, and door gasket, regularly to remove any limescale deposits that have accumulated. Vinegar or citric acid can be used as natural cleaners to dissolve mineral deposits.

Descaling Solution: Periodically run a descaling cycle using a commercial descaling solution or a homemade solution of vinegar and water to remove limescale deposits from dishwasher components.

Inspect and Maintain Components: Regularly inspect dishwasher components such as the heating element, spray arms, and filter for signs of limescale buildup or damage. Clean or replace components as needed to maintain optimal performance.

Adjust Detergent Usage: Adjust the amount of detergent used in the dishwasher according to the hardness of the water in your area. Using too much detergent can contribute to the buildup of soap scum and limescale deposits.

Electric Water Heater

NO HOT WATER

When an electric water heater fails to produce hot water, several common issues may be the cause:

Tripped Circuit Breaker: Check the circuit breaker or fuse box to ensure that power is being supplied to the water heater. A tripped breaker or blown fuse can disrupt the heating process.

Faulty Heating Element: Electric water heaters typically have one or two heating elements, and if one or both of them fail, the water will not heat properly. Testing the heating elements for continuity using a multimeter can determine if they are functioning correctly.

Broken Thermostat: The thermostat controls the temperature of the water in the tank. If it malfunctions or is out of calibration, it may not signal the heating elements to activate, resulting in no hot water.

Sediment Buildup: Over time, sediment from the water supply can accumulate at the bottom of the tank, insulating the heating elements and reducing their effectiveness. Flushing the tank to remove sediment buildup can restore proper heating.

Faulty Wiring or Connections: Loose or damaged wiring or connections within the water heater can interrupt the flow of electricity to the heating elements, preventing them from heating the water.

High Demand: If there is an unusually high demand for hot water, such as multiple showers or loads of laundry running simultaneously, the water heater may struggle to keep up with the demand, resulting in lukewarm or cold water.

Thermostat Settings: Ensure that the thermostat on the water heater is set to the desired temperature. If it is set too low, the water may not reach the desired temperature.

Insufficient Tank Size: If the water heater tank is too small for the household's hot water needs, it may run out of hot water quickly, especially during periods of high demand.

To address the issue of not having hot water from an electric water heater:

Check Power Supply: Verify that the water heater is receiving power by checking the circuit breaker or fuse box.

Inspect Heating Elements: Test the heating elements for continuity using a multimeter. If one or both elements fail the test, they may need to be replaced.

Check Thermostat: Test the thermostat for proper function and calibration. If it is faulty, it may need to be replaced.

Flush the Tank: Drain and flush the water heater tank to remove sediment buildup, which can improve heating efficiency.

Check Wiring and Connections: Inspect the wiring and connections within the water heater for any damage or looseness. Repair or replace any faulty components as needed.

Adjust Thermostat Settings: Ensure that the thermostat is set to the desired temperature. Adjust it if necessary to achieve the desired hot water temperature.

Consider Tank Size: If the water heater tank is consistently running out of hot water, consider upgrading to a larger tank or a more efficient model to meet the household's hot water needs.

INSUFFICIENT HOT WATER

When an electric water heater produces insufficient hot water, it can be frustrating and inconvenient. Here are common issues that may cause this problem:

Undersized Water Heater: If the water heater is too small for the household's hot water demand, it may struggle to keep up, resulting in insufficient hot water. Consider upgrading to a larger-capacity water heater if this is the case.

Sediment Buildup: Over time, sediment from the water supply can accumulate at the bottom of the tank, reducing the amount of available hot water and lowering efficiency. Flushing the tank can help remove sediment buildup and restore hot water production.

Faulty Heating Elements: Electric water heaters typically have one or two heating elements that heat the water. If one or both of these elements are faulty or burned out, the water heater may not produce enough hot water. Testing and replacing the heating elements can resolve this issue.

Incorrect Thermostat Settings: If the thermostat on the water heater is set too low, the water may not reach the desired temperature. Adjusting the thermostat to a higher temperature can help increase hot water production.

Dip Tube Issues: The dip tube is a component that delivers cold water to the bottom of the tank for heating. If the dip tube is broken or deteriorated, it may not effectively distribute cold water, resulting in insufficient hot water at the taps.

High Demand Periods: During periods of high demand for hot water, such as multiple showers or laundry loads running simultaneously, the water heater may struggle to keep up. This can lead to temporary shortages of hot water.

Faulty Thermostat: A malfunctioning thermostat may not accurately control the temperature of the water heater, resulting in insufficient hot water production. Testing and replacing the thermostat can address this issue.

Power Supply Issues: Problems with the electrical supply to the water heater, such as a tripped circuit breaker or loose wiring, can interrupt heating and result in insufficient hot water.

Old Age: Over time, the efficiency of the water heater can decline due to wear and tear. If the water heater is old and showing signs of deterioration, it may be time for a replacement.

To address insufficient hot water in an electric water heater:

Check Thermostat Settings: Ensure that the thermostat on the water heater is set to the desired temperature and adjust it if necessary.

Flush the Tank: Periodically flush the water heater tank to remove sediment buildup and improve heating efficiency.

Test Heating Elements: Use a multimeter to test the heating elements for continuity and replace them if they are faulty.

Inspect Dip Tube: Inspect the dip tube for signs of damage or deterioration and replace it if necessary.

Check Power Supply: Verify that the water heater is receiving power and that there are no issues with the electrical supply.

Consider Water Usage Habits: Adjusting water usage habits, such as spacing out showers or laundry loads, can help reduce hot water demand during peak periods.

Upgrade the Water Heater: If the water heater is old or undersized, consider upgrading to a larger capacity or more efficient model to meet the household's hot water needs.

WATER LEAKS

Water leaks in an electric water heater can lead to various problems, including damage to the surrounding area, increased energy bills, and potential safety hazards. Common causes of water leaks in electric water heaters include:

Faulty Pressure Relief Valve (PRV): The pressure relief valve is designed to release excess pressure from the tank to prevent it from bursting. A leaking PRV may indicate high pressure in the tank or a faulty valve that needs replacement.

Leaking Drain Valve: The drain valve is located at the bottom of the water heater and is used for flushing sediment from the tank. A leaking drain valve can occur due to wear, corrosion, or improper closure after flushing.

Corroded or Leaking Tank: Over time, the tank itself may corrode, leading to leaks. Corrosion can be accelerated in areas with hard water. If the tank is leaking, it often signals the need for a replacement.

Faulty Temperature and Pressure (T&P) Relief Valve: The T&P relief valve is responsible for releasing pressure and temperature buildup. A leaking T&P valve may indicate excessive pressure or temperature, or the valve itself may be faulty and require replacement.

Loose or Leaky Pipe Connections: Water heaters have various pipes and connections that may become loose or develop leaks over time. Check all connections, including inlet and outlet pipes, for signs of water leakage.

Condensation: In humid conditions, condensation may form on the exterior of the water heater, giving the appearance of a leak. Confirm that the source of the water is not condensation before addressing other potential issues.

To address water leaks in an electric water heater:

Shut Off the Power: Before inspecting or attempting any repairs, turn off the power supply to the water heater by switching off the circuit breaker.

Check the Pressure Relief Valve: Test the pressure relief valve by lifting the lever to release some water. If the valve continues to leak after releasing the lever, it may be faulty and should be replaced.

Inspect the Drain Valve: Check the drain valve for leaks. If it is leaking, consider tightening the valve or replacing it if the leak persists.

Examine the Tank: Inspect the tank for any signs of corrosion or visible leaks. If the tank itself is leaking, it may be necessary to replace the water heater.

Inspect Pipe Connections: Check all pipe connections for signs of leaks. Tighten any loose connections and replace any damaged or corroded pipes.

Replace Faulty T&P Relief Valve: If the T&P relief valve is leaking, it may need to be replaced. Ensure that the replacement valve has the appropriate pressure and temperature ratings.

Address Condensation Issues: If the water heater is located in a humid environment, consider reducing humidity in the area or insulating the tank to minimize condensation.

STRANGE NOISE

Strange noises coming from an electric water heater can be indicative of various issues within the appliance. Some common causes of these noises include:

Sediment Buildup: Sediment accumulates at the bottom of the water heater tank over time, especially in areas with hard water. During heating cycles, this sediment can harden and create popping, banging, or rumbling noises as it is disturbed by the heating elements.

Expansion and Contraction: As the water heater heats and cools, the metal tank expands and contracts. This process can produce popping or creaking sounds, particularly if the tank is old or improperly insulated.

Mineral Deposits: Mineral deposits can form on the heating elements or inside the tank, leading to hissing, sizzling, or popping noises as water interacts with the buildup during heating cycles.

Loose Heating Elements: If the heating elements inside the tank are loose or not properly secured, they may vibrate or move during operation, causing knocking or rattling sounds.

Faulty Dip Tube: A damaged or deteriorated dip tube can cause noises as it distributes cold water to the bottom of the tank. This issue may result in water not heating properly or inconsistent water temperature.

Water Hammer: Water hammer occurs when water flow is suddenly stopped or redirected, causing pipes to vibrate or bang against surrounding surfaces. This can produce loud, repetitive thumping or banging noises.

High Water Pressure: Excessive water pressure can lead to noisy operation of the water heater, particularly when valves or other components open or close. This issue may require the installation of a pressure-reducing valve.

Loose or Faulty Components: Loose connections, damaged valves, or faulty components within the water heater can cause vibrations or unusual sounds during operation.

To address strange noises in an electric water heater:

Flush the Tank: Regularly flushing the water heater tank to remove sediment buildup can help reduce popping, banging, or rumbling noises caused by sediment disturbance during heating cycles.

Inspect Heating Elements: Check the heating elements for signs of damage or corrosion. Tighten loose elements and consider replacing them if they are worn out or faulty.

Replace Dip Tube: If the dip tube is damaged or deteriorated, replace it to ensure proper distribution of cold water within the tank.

Install Water Hammer Arrestors: Water hammer arrestors can be installed on the water supply lines to reduce the occurrence of water hammer and associated noises.

Adjust Water Pressure: If the water pressure is too high, install a pressure-reducing valve to bring it within the recommended range and minimize noise.

Inspect and Tighten Connections: Check all connections, valves, and components within the water heater for signs of looseness or damage. Tighten or replace any faulty components as needed.

Consider Insulation: Adding insulation to the water heater tank or pipes can help reduce expansion and contraction noises caused by temperature changes.

RUSTY AND DISCOLORED WATER

Rusty or discolored water coming from an electric water heater can be a cause for concern and may indicate various issues within the appliance or plumbing system. Common causes of rusty or discolored water in an electric water heater include:

Corroded Anode Rod: The sacrificial anode rod is designed to attract corrosive elements in the water and protect the tank from corrosion. Over time, the anode rod can become depleted and corroded, leading to rusty or discolored water.

Corroded Tank: If the interior lining of the water heater tank has deteriorated or if the tank itself is corroded, rust particles may contaminate the water, causing discoloration.

Sediment Buildup: Sediment from the water supply can accumulate at the bottom of the water heater tank over time, especially in areas with hard water. When disturbed, this sediment can cause water discoloration and may contain rust particles.

Galvanized Piping: If the plumbing system contains galvanized pipes, corrosion and rust buildup within the pipes can cause rusty or discolored water throughout the home, including at the hot water taps supplied by the water heater.

Corroded Pipe Connections: Corrosion or rust buildup at pipe connections or fittings connected to the water heater can introduce rusty or discolored water into the plumbing system.

Water Main Break or Disturbance: A break or disturbance in the water main serving the home can introduce sediment or rust into the water supply, leading to temporary discoloration.

High Iron Content in Water: Some water sources have naturally high levels of iron, which can cause water discoloration, especially when heated.

To address rusty or discolored water in an electric water heater:

Flush the Tank: Flushing the water heater tank can help remove sediment buildup and rust particles that may be contributing to water discoloration. Follow the manufacturer's instructions for flushing the tank safely.

Inspect and Replace Anode Rod: Check the condition of the sacrificial anode rod and replace it if it is corroded or depleted. A new anode rod can help protect the tank from further corrosion.

Check for Tank Corrosion: Inspect the interior of the water heater tank for signs of corrosion or rust. If the tank is corroded, it may need to be replaced to prevent further contamination of the water supply.

Flush Plumbing System: If rusty or discolored water is present throughout the home, flush the plumbing system by running cold water taps until the water runs clear.

Test Water Quality: Consider testing the water quality to determine if high iron levels or other contaminants are present. Water treatment solutions may be necessary to address specific water quality issues.

Consider Plumbing Upgrades: If the plumbing system contains galvanized pipes or corroded fittings, consider upgrading to more durable materials such as copper or PEX to prevent future rust and corrosion issues.

HIGH ENERGY BILL

A high energy bill associated with an electric water heater can be a result of various factors related to inefficiency or excessive energy consumption. Common issues contributing to a high energy bill with an electric water heater include:

Old or Inefficient Water Heater: Aging water heaters may become less efficient over time, resulting in increased energy consumption. If the water heater is outdated or has not been properly maintained, it may be less energy efficient than newer models.

Inadequate Insulation: Insufficient insulation around the water heater tank or hot water pipes can lead to heat loss, requiring the water heater to work harder and consume more energy to maintain the desired water temperature.

High Water Temperature Setting: Setting the water heater thermostat to a higher temperature than necessary can result in excessive energy consumption. Lowering the thermostat setting to the recommended temperature (typically around 120°F or 49°C) can help reduce energy usage without sacrificing comfort.

Frequent Use or High Demand: If there is a high demand for hot water in the household, such as multiple showers, baths, or laundry loads in quick succession, the water heater may run frequently, consuming more energy.

Sediment Buildup: Accumulation of sediment at the bottom of the water heater tank can insulate the heating elements and reduce heating efficiency. This can result in longer heating cycles and increased energy consumption.

Faulty Heating Elements: Malfunctioning or damaged heating elements may not heat water efficiently, leading to increased energy consumption as the water heater works harder to maintain the desired temperature.

Leaking Hot Water: A leaking hot water faucet or plumbing fixture can result in wasted hot water and increased energy usage to heat replacement water.

Improperly Sized Water Heater: If the water heater is undersized for the household's hot water needs, it may struggle to keep up with demand, leading to frequent heating cycles and increased energy consumption.

To address a high energy bill associated with an electric water heater:

Insulate the Water Heater: Adding insulation blankets or jackets to the water heater tank and insulating hot water pipes can help reduce heat loss and improve energy efficiency.

Lower the Thermostat Setting: Adjust the water heater thermostat to the recommended temperature setting to minimize energy consumption while still providing comfortable hot water.

Schedule Regular Maintenance: Periodically flush the water heater tank to remove sediment buildup and ensure optimal heating efficiency. Replace damaged or malfunctioning heating elements as needed.

Address Hot Water Leaks: Repair any leaking hot water faucets or fixtures promptly to prevent wasted hot water and excessive energy usage.

Consider Energy-Efficient Upgrades: If the water heater is old or inefficient, consider upgrading to a more energy-efficient model with features such as improved insulation, high-efficiency heating elements, or heat pump technology.

Monitor Hot Water Usage: Encourage water conservation practices and avoid unnecessary hot water usage to reduce energy consumption.

PRV ISSUES

The pressure relief valve (PRV) in an electric water heater is a crucial safety component designed to release excess pressure from the tank to prevent it from bursting. Common issues with the PRV in an electric water heater include:

Leaking: A leaking PRV is a common issue and can be caused by several factors, including high water pressure, a faulty valve, or debris obstructing the valve seat. A leaking PRV should be addressed promptly to prevent water wastage and potential damage to the water heater.

Stuck or Jammed Valve: Sometimes the PRV may become stuck or jammed due to mineral buildup, corrosion, or debris. A stuck valve may fail to operate properly in an overpressure situation, posing a safety risk. Regular testing and maintenance of the PRV can help prevent this issue.

Failure to Open: In some cases, the PRV may fail to open during an overpressure event, leading to potential tank rupture. This failure can occur due to valve malfunction, corrosion, or obstruction of the valve seat. Proper installation and periodic testing of the PRV are essential to ensure it functions correctly.

Continuous Discharge: If the PRV continuously discharges water, it may indicate an overpressure condition in the water heater tank. This could be due to high water pressure from the supply line, thermal expansion of water, or a malfunctioning thermostat. Addressing the root cause of the overpressure situation is essential to prevent further PRV discharge.

Inadequate Pressure Relief: In some cases, the PRV may not relieve pressure adequately, leading to potential safety hazards. This can occur due to valve malfunction, improper installation, or incorrect sizing of the PRV. Installing a properly sized PRV and ensuring it is correctly installed and maintained can help prevent this issue.

Corrosion and Degradation: Over time, the PRV may degrade due to corrosion, mineral buildup, or wear and tear. Regular inspection and replacement of the PRV as needed can help ensure it remains in good working condition.

Improper Installation: Incorrect installation of the PRV, such as improper seating or inadequate sealing, can lead to performance issues and safety concerns. It's essential to follow the manufacturer's instructions and industry best practices when installing or replacing the PRV.

To address PRV issues in an electric water heater:

Inspect for Leaks: Regularly inspect the PRV for any signs of leaks or discharge. Address any leaks promptly by repairing or replacing the PRV as needed.

Test the PRV: Periodically test the PRV to ensure it operates correctly. Follow the manufacturer's instructions for testing and maintenance procedures.

Check Water Pressure: Monitor water pressure from the supply line to ensure it remains within the recommended range. Installing a pressure regulator can help control high water pressure and prevent PRV discharge.

Flush the Tank: Periodically flush the water heater tank to remove sediment and debris that could obstruct the PRV or valve seat.

Replace as Needed: If the PRV shows signs of corrosion, degradation, or malfunction, replace it with a new PRV of the appropriate size and rating.

SHORT CYCLING

Short cycling in an electric water heater refers to the frequent turning on and off of the heating elements, which can result in inefficient operation, increased energy consumption, and premature wear on the appliance. Common issues that can cause short cycling in an electric water heater include:

Thermostat Setting: If the thermostat is set too high, the water heater may reach the set temperature quickly and shut off prematurely, leading to short cycling. Adjusting the thermostat to the recommended temperature setting can help prevent this issue.

Sediment Buildup: Accumulation of sediment at the bottom of the water heater tank can insulate the heating elements, causing them to overheat and cycle on and off frequently. Flushing the tank to remove sediment buildup can improve heating efficiency and reduce short cycling.

Faulty Thermostat: A malfunctioning thermostat may inaccurately detect water temperature, causing the heating elements to cycle on and off unnecessarily. Testing and replacing the thermostat as needed can resolve this issue.

Incorrect Wiring or Connections: Loose or faulty wiring or connections within the water heater can disrupt the electrical supply to the heating elements, leading to short cycling. Inspecting and repairing wiring and connections can help prevent this issue.

High Water Temperature: If the water temperature in the tank exceeds the set limit, the thermostat may shut off the heating elements prematurely, resulting in short cycling. Adjusting the thermostat setting or installing a temperature and pressure relief valve can help regulate water temperature and prevent overheating.

Insufficient Insulation: Inadequate insulation around the water heater tank can cause heat loss, triggering the thermostat to cycle on and off more frequently to maintain the desired temperature. Adding insulation to the tank can help improve efficiency and reduce short cycling.

Faulty Heating Elements: Malfunctioning or damaged heating elements may not heat water efficiently, leading to short cycling as the water heater struggles to maintain the desired temperature. Testing and replacing faulty heating elements can address this issue.

High Water Pressure: Excessive water pressure in the plumbing system can cause the pressure relief valve to open, resulting in short cycling as the water heater refills and reheats the tank. Installing a pressure regulator can help stabilize water pressure and prevent short cycling.

To address short cycling in an electric water heater:

Adjust Thermostat Setting: Set the thermostat to the recommended temperature setting to prevent overheating and premature shutoff.

Flush the Tank: Periodically flush the water heater tank to remove sediment buildup and improve heating efficiency.

Test and Replace Thermostat: Check the thermostat for proper function and replace it if it is faulty or inaccurate.

Inspect Wiring and Connections: Inspect and repair any loose or faulty wiring or connections within the water heater.

Install Insulation: Add insulation to the water heater tank to reduce heat loss and improve efficiency.

Test and Replace Heating Elements: Test the heating elements for proper function and replace them if they are faulty or damaged.

Address High Water Pressure: Install a pressure regulator to stabilize water pressure and prevent pressure-related short cycling.

COLD WATER SANDWICH

Cold water sandwich is a phenomenon that occurs in certain types of water heaters, including electric models, where the user experiences a sudden burst of cold water sandwiched between hot water during usage. This issue can be frustrating and uncomfortable. Common causes of cold water sandwich in an electric water heater include:

Inadequate Temperature Regulation: If the water heater's thermostat is not accurately regulating the temperature, it can lead to sudden fluctuations in water temperature, resulting in a cold water sandwich effect.

Improperly Sized Water Heater: Water heaters that are undersized for the household's hot water demand may struggle to maintain a consistent temperature, leading to temperature fluctuations and cold water sandwich.

Thermal Stratification: In large water heaters, such as those with high storage capacities, layers of hot and cold water can form within the tank, leading to temperature variations at the outlet. When hot water is drawn from the tank, cold water from the lower portion may mix with the hot water, causing the cold water sandwich effect.

Cross-Connections or Mixing: Cross-connections between hot and cold water lines or mixing within the plumbing system can result in cold water entering the hot water line, causing intermittent bursts of cold water during usage.

Flow Rate Variations: Variations in water flow rate, such as changes in water pressure or usage patterns, can exacerbate the cold water sandwich effect by causing an uneven mixing of hot and cold water within the plumbing system.

To address cold water sandwich in an electric water heater:

Adjust Temperature Settings: Ensure that the water heater's thermostat is set to the recommended temperature setting (typically around 120°F or 49°C) to minimize temperature fluctuations.

Install Temperature-Regulating Valves: Temperature-regulating valves, such as thermostatic mixing valves, can help mitigate temperature variations by blending hot and cold water to maintain a consistent temperature at the outlet.

Proper Sizing: Ensure that the water heater is appropriately sized to meet the household's hot water demand. An undersized water heater may struggle to maintain a consistent temperature, leading to cold water sandwich.

Thermal Mixing Solutions: Installing devices such as recirculation pumps or thermal mixing valves can help prevent thermal stratification within the water heater tank, reducing the likelihood of cold water sandwich.

Check for Cross-Connections: Inspect the plumbing system for any cross-connections between hot and cold water lines and correct any issues found.

Monitor and Adjust Flow Rates: Monitor water flow rates and adjust pressure-reducing valves or flow restrictors as needed to maintain consistent water pressure and flow throughout the plumbing system.

CORROSION AND RUST

Corrosion and rust in electric water heaters can lead to various issues, including reduced efficiency, damage to the tank, and potential water quality concerns. Common causes and issues related to corrosion and rust in electric water heaters include:

Sacrificial Anode Rod Depletion: The sacrificial anode rod inside the water heater tank is designed to attract corrosive elements and protect the tank from rusting. Over time, the anode rod can become depleted, leading to corrosion of the tank itself.

Solution: Regularly inspect and replace the sacrificial anode rod as needed to maintain effective corrosion protection.

Tank Corrosion: If the anode rod is not replaced in a timely manner or if the water heater is exposed to aggressive water conditions, the tank itself may corrode. Corrosion can lead to leaks and structural damage.

Solution: If the tank is corroded, it may be necessary to replace the water heater. Regular maintenance and anode rod replacement can help prevent tank corrosion.

Mineral Buildup and Sediment: Sediment and mineral buildup at the bottom of the tank can contribute to corrosion. When sediment accumulates, it can create a barrier between the water and the anode rod, allowing corrosion to occur.

Solution: Periodically flush the water heater tank to remove sediment and mineral buildup. This helps maintain the effectiveness of the anode rod and reduces the risk of corrosion.

Galvanic Corrosion: Galvanic corrosion can occur when different metals are in contact within the water heater. This can happen if there are dissimilar metals present, such as copper pipes connected to a steel tank.

Solution: Use dielectric unions or other measures to prevent direct contact between dissimilar metals and minimize the risk of galvanic corrosion.

High Chloride Levels in Water: High chloride levels in water can accelerate corrosion. This is often more prevalent in homes with well water or in regions with specific water chemistry.

Solution: If chloride levels are high, consider water treatment options or consult with a water quality professional to address the issue.

Excessive Water Temperature: Operating the water heater at excessively high temperatures can accelerate corrosion. Higher temperatures can cause more aggressive corrosion reactions.

Solution: Set the water heater thermostat to the recommended temperature (typically around 120°F or 49°C) to reduce the risk of corrosion.

External Factors: External factors such as exposure to corrosive chemicals or environmental conditions can contribute to rust and corrosion.

Solution: Protect the water heater from exposure to corrosive substances and ensure that it is installed in a suitable environment.

Gas Water Heater

PILOT LIGHT PROBLEMS

Pilot light problems are a common issue with gas water heaters and can result in no hot water. Some common causes of pilot light problems include:

Thermocouple Issues: The thermocouple is a safety device that detects the presence of the pilot light flame. If the thermocouple is faulty or dirty, it may not detect the flame, causing the gas valve to shut off and the pilot light to go out.

Dirty Pilot Assembly: Dirt, dust, or corrosion on the pilot assembly can obstruct the flow of gas to the pilot light or interfere with the flame. Cleaning the pilot assembly can often resolve this issue.

Gas Supply Issues: A disruption in the gas supply, such as a closed gas valve or a gas line leak, can prevent the pilot light from igniting or cause it to go out unexpectedly.

Airflow Problems: Poor ventilation or a draft in the area around the water heater can affect the pilot light's ability to stay lit. Make sure the area around the water heater is well ventilated and free from obstructions.

Faulty Gas Valve: A malfunctioning gas valve may not supply gas to the pilot light properly, preventing it from igniting or causing it to go out unexpectedly.

Thermostat Problems: Issues with the water heater's thermostat, such as misalignment or malfunction, can affect the pilot light's operation and cause it to go out.

Pilot Orifice Blockage: The pilot orifice, which controls the flow of gas to the pilot light, may become blocked by debris or corrosion, preventing the pilot light from igniting.

To troubleshoot and address pilot light problems:

Check Gas Supply: Ensure that the gas supply valve is open and that there are no leaks in the gas line. If the gas supply is interrupted, the pilot light will not ignite or stay lit.

Inspect Thermocouple: Check the thermocouple for proper positioning and alignment with the pilot flame. If the thermocouple is dirty or corroded, clean it with a soft brush or sandpaper.

Clean Pilot Assembly: Remove any dirt, dust, or corrosion from the pilot assembly using compressed air or a soft brush. Make sure the pilot orifice is clear of any obstructions.

Check Ventilation: Ensure that the area around the water heater is well ventilated and free from obstructions that could affect airflow to the pilot light.

Inspect Gas Valve: Test the gas valve for proper operation and ensure that it is supplying gas to the pilot light when it is supposed to.

Adjust Thermostat: Check the thermostat settings and make sure they are correctly adjusted. If the thermostat is malfunctioning, consider replacing it.

BURNER PROBLEMS

Burner issues in a gas water heater can lead to inadequate heating or no hot water. Common problems with the burner assembly include:

Clogged Burner Orifice: Dirt, dust, or debris can accumulate in the burner orifice, obstructing the flow of gas and affecting the flame. A clogged burner orifice can result in uneven heating or no hot water.

Dirty Burner Assembly: The burner assembly may become dirty or corroded over time, hindering the proper combustion of gas and affecting the flame quality. Cleaning the burner assembly can often resolve this issue.

Misaligned Burner: If the burner is not properly aligned or positioned within the combustion chamber, it may not receive the correct airflow or gas supply, resulting in inefficient combustion and poor heating performance.

Thermocouple Malfunction: The thermocouple is a safety device that detects the presence of the pilot light flame. If the thermocouple is faulty or malfunctioning, it may not signal the gas valve to open, preventing the burner from igniting.

Gas Valve Problems: A malfunctioning gas valve may not supply gas to the burner properly, preventing it from igniting or causing it to shut off unexpectedly.

Ventilation Issues: Poor ventilation or inadequate airflow around the burner can affect combustion and flame quality. Make sure the area around the water heater is well ventilated and free from obstructions.

Gas Pressure Problems: Insufficient gas pressure from the supply line can prevent the burner from igniting or cause it to produce a weak flame. Checking the gas pressure and adjusting it as needed can help resolve this issue.

Ignition Problems: If the ignition system, such as the spark igniter or pilot light, is faulty or malfunctioning, the burner may not ignite properly. Checking and repairing the ignition system can help restore proper burner operation.

To address burner issues in a gas water heater:

Inspect and Clean Burner Assembly: Remove any dirt, dust, or debris from the burner assembly using compressed air or a soft brush. Ensure that the burner orifice is clear and free from obstructions.

Check Thermocouple: Test the thermocouple for proper operation and ensure that it is detecting the pilot light flame correctly. Replace the thermocouple if it is faulty or malfunctioning.

Inspect Gas Valve: Test the gas valve for proper operation and ensure that it is supplying gas to the burner when it is supposed to. Replace the gas valve if it is faulty or malfunctioning.

Adjust Ventilation: Ensure that the area around the water heater is well ventilated and free from obstructions that could affect airflow to the burner.

Check Gas Pressure: Verify that the gas pressure from the supply line is within the recommended range. Adjust the gas pressure if necessary to ensure proper burner operation.

Test Ignition System: Check the ignition system, including the spark igniter or pilot light, for proper operation. Repair or replace any faulty components as needed.

GAS VALVE ISSUES

Gas valve issues in a gas water heater can lead to problems with ignition, heating, or safety concerns. Here are some common gas valve issues:

Faulty Gas Valve Solenoid: The solenoid is an electromechanical component responsible for opening and closing the gas valve. If the solenoid fails, it may not allow gas to flow to the burner, preventing ignition and heating.

Clogged Gas Valve Orifice: Dirt, dust, or debris can accumulate in the gas valve orifice, obstructing the flow of gas to the burner. A clogged orifice can result in inadequate heating or no hot water.

Gas Leak: A leak in the gas valve or associated fittings can pose a safety hazard and result in gas odor or potential ignition issues. Gas leaks should be addressed immediately by a qualified professional.

Faulty Gas Valve Control Board: The gas valve control board regulates the operation of the gas valve and may malfunction due to electrical issues or component failure. A faulty control board can prevent the gas valve from opening or closing properly.

Stuck Gas Valve: The gas valve may become stuck in the closed position due to corrosion, debris, or mechanical issues. A stuck gas valve will prevent gas flow to the burner, resulting in no heat.

Gas Pressure Regulator Problems: The gas pressure regulator regulates the gas pressure supplied to the water heater. If the regulator malfunctions, it can result in improper gas pressure, affecting burner operation and heating performance.

Thermocouple Issues: The thermocouple is a safety device that detects the presence of the pilot light flame. If the thermocouple is faulty or malfunctioning, it may not signal the gas valve to open, preventing ignition and heating.

To address gas valve issues in a gas water heater:

Inspect for Gas Leaks: Check for gas leaks using a gas leak detector solution or by smelling for gas odors. If a gas leak is detected, shut off the gas supply to the water heater immediately and contact a qualified professional for repair.

Clean Gas Valve Orifice: Remove any dirt, dust, or debris from the gas valve orifice using compressed air or a soft brush. Ensure that the orifice is clear and free from obstructions.

Check Gas Pressure: Verify that the gas pressure supplied to the water heater is within the recommended range. Adjust the gas pressure regulator if necessary to ensure proper burner operation.

Test Solenoid and Control Board: Test the gas valve solenoid and control board for proper operation using a multimeter or by consulting the manufacturer's guidelines. Replace any faulty components as needed.

Inspect Thermocouple: Test the thermocouple for proper operation and ensure that it is detecting the pilot light flame correctly. Replace the thermocouple if it is faulty or malfunctioning.

Ensure Proper Ventilation: Ensure that the area around the water heater is well ventilated and free from obstructions that could affect gas combustion or flow.

THERMOCOUPLE FAILURE

Thermocouple failure in a gas water heater can lead to issues with the pilot light, burner, or gas valve operation. Some common causes of thermocouple failure include:

Pilot Light Problems: If the pilot light goes out frequently due to issues such as a dirty or misaligned pilot assembly, improper gas pressure, or drafts, the thermocouple may be exposed to excessive heat or insufficient flame, leading to premature failure.

Thermocouple Misalignment: Improper positioning or misalignment of the thermocouple relative to the pilot flame can prevent it from accurately detecting the flame, causing it to fail prematurely.

Corrosion or Contamination: Corrosion, rust, or contamination on the thermocouple surface or connections can interfere with its ability to generate a voltage signal in response to the pilot flame, leading to failure.

Overheating: Excessive heat exposure from nearby sources, such as the burner or combustion chamber, can cause the thermocouple to overheat and degrade over time, resulting in failure.

Wiring Issues: Loose or damaged wiring connections between the thermocouple and the gas valve or control board can disrupt the flow of electrical signals, leading to thermocouple failure.

Age and Wear: Like any mechanical component, thermocouples have a limited lifespan and may wear out over time due to normal usage and aging. Regular maintenance and replacement may be necessary to prevent failure.

Gas Valve Problems: Malfunctioning gas valves can prevent the proper operation of the pilot light or interfere with the thermocouple's ability to detect the flame, leading to thermocouple failure.

Repairing a thermocouple in a gas water heater involves several steps. Here's a general guide:

Turn Off Gas Supply: Shut off the gas supply to the water heater by turning the gas valve to the "off" position.

Allow Cooling: Allow the water heater to cool down for at least thirty minutes before proceeding with any repairs to avoid burns or injuries.

Locate Thermocouple: The thermocouple is a small metal rod located near the pilot light assembly. It typically extends from the gas control valve to the pilot light.

Disconnect Thermocouple: Use a wrench to disconnect the thermocouple from the gas control valve. Be gentle to avoid damaging any components.

Remove Pilot Assembly: Depending on the water heater model, you may need to remove the pilot assembly to access the thermocouple fully. Follow the manufacturer's instructions for disassembly.

Inspect Thermocouple: Check the thermocouple for signs of damage, corrosion, or wear. Look for any visible defects that may indicate the need for replacement.

Clean Thermocouple: If the thermocouple is dirty or corroded, gently clean it using fine-grit sandpaper or a soft cloth. Ensure that the surface is smooth and free from debris.

Reassemble Pilot Assembly: If you removed the pilot assembly, reassemble it according to the manufacturer's instructions. Make sure all components are properly aligned and secured.

Reconnect Thermocouple: Reconnect the thermocouple to the gas control valve and tighten the connection securely using a wrench.

Test Pilot Light: Turn on the gas supply and relight the pilot light according to the manufacturer's instructions. Hold down the pilot light button or knob for a few seconds to allow the thermocouple to heat up and generate a signal.

Check Operation: Once the pilot light is lit, observe its flame. The flame should be steady and blue, with the thermocouple positioned in the flame. If the pilot light remains lit without flickering, the repair is successful.

Monitor for Issues: Keep an eye on the pilot light over the next few days to ensure that it remains lit consistently.

SEDIMENT BUILDUP

Sediment buildup is a common issue in gas water heaters and can lead to various problems, including reduced efficiency, increased energy consumption, and potential damage to the tank. Several factors contribute to sediment buildup in a gas water heater:

Hard Water: Hard water contains high levels of minerals such as calcium and magnesium. When hard water is heated in the water heater tank, these minerals can precipitate out of solution and settle at the bottom of the tank, forming sediment.

Lack of Maintenance: Neglecting regular maintenance tasks such as flushing the water heater tank can allow sediment to accumulate over time. Without periodic removal, sediment buildup can become significant and impair the water heater's performance.

Old Age: As water heaters age, they are more prone to sediment buildup due to wear and corrosion within the tank. Older water heaters may have accumulated more sediment over time, leading to decreased efficiency and increased energy consumption.

High Water Temperature: Operating the water heater at a higher temperature can accelerate mineral precipitation and sediment buildup in the tank. Higher temperatures cause minerals to precipitate out of solution more rapidly, leading to more significant sediment accumulation.

Insufficient Anode Rod Protection: The sacrificial anode rod inside the water heater tank helps protect the tank from corrosion and sediment buildup by attracting corrosive elements. If the anode rod becomes depleted or corroded, it may not provide adequate protection against sediment accumulation.

Poor Water Quality: Water with high levels of suspended solids or contaminants can contribute to sediment buildup in the water heater tank. Impurities in the water can settle at the bottom of the tank over time, forming sediment.

To prevent sediment buildup in a gas water heater:

Regular Maintenance: Perform annual maintenance tasks such as flushing the water heater tank to remove sediment buildup. Flushing the tank involves draining a portion of the water to remove accumulated sediment.

Install a Water Softener: Consider installing a water softener to treat hard water and reduce mineral deposits in the water heater tank. Water softeners remove calcium and magnesium ions from the water, reducing the likelihood of sediment buildup.

Monitor Anode Rod: Inspect the sacrificial anode rod periodically and replace it if it is depleted or corroded. A functioning anode rod helps protect the tank from corrosion and sediment buildup.

Lower Water Temperature: Lowering the water heater's temperature setting can reduce mineral precipitation and sediment buildup in the tank. Set the temperature to a moderate level to minimize sediment accumulation.

Address Water Quality Issues: If your water supply has high levels of impurities or contaminants, consider installing a filtration system or water treatment device to improve water quality and reduce sediment buildup.

By addressing these common issues and implementing preventive measures, you can reduce sediment buildup in your gas water heater and maintain its efficiency and performance over time. Regular maintenance and monitoring of the water heater can help identify and address sediment buildup before it becomes a significant problem. If you are unsure how to perform maintenance tasks or if sediment buildup persists despite your efforts, consult a qualified technician or plumber for assistance.

PRV MALFUNCTIONS

Pressure relief valve (PRV) malfunctions in gas water heaters can pose safety risks and lead to issues such as leaks or excessive pressure buildup. Several common causes of PRV malfunctions include:

Excessive Pressure: If the water pressure in the tank exceeds safe levels, the PRV may activate to release the excess pressure. Continuous activation of the PRV can cause wear and eventual malfunction.

Sediment Buildup: Sediment accumulation in the water heater tank can obstruct the PRV's valve mechanism, preventing it from operating correctly. Sediment buildup may also lead to overheating and pressure fluctuations, triggering the PRV unnecessarily.

Corrosion: Corrosion can affect the PRV's components, including the valve and spring, leading to stiffness or failure to open properly. Corrosion may occur due to exposure to moisture or corrosive elements in the water supply.

Age and Wear: Over time, PRVs may experience wear and deterioration of internal components, reducing their effectiveness and reliability. Aging PRVs may become less responsive or fail to operate as intended.

Improper Installation: Incorrect installation or sizing of the PRV can affect its performance and reliability. For example, installing a PRV with an incorrect pressure rating or orientation can lead to malfunction.

High Temperature: Elevated temperatures within the water heater tank can affect the PRV's components and cause them to degrade over time. High temperatures may soften seals, deform components, or impair the PRV's ability to function correctly.

Water Hammer: Water hammer occurs when abrupt changes in water flow or pressure cause shockwaves in the plumbing system. Repeated water hammer events can damage the PRV and lead to malfunction or failure.

Repairing a pressure relief valve (PRV) malfunction in a gas water heater typically involves several steps. Here's a general guide to repairing a PRV issue:

Safety First: Before beginning any repair work, ensure the gas supply to the water heater is turned off. Additionally, allow the water heater to cool down to a safe temperature to prevent burns.

Locate the PRV: The PRV is typically located on or near the top of the water heater tank. It's a valve with a lever or handle and a discharge pipe connected to it.

Inspect for Damage: Visually inspect the PRV for any signs of damage, such as corrosion, leaks, or visible wear on the valve or surrounding components. If the PRV is damaged, it will likely need to be replaced rather than repaired.

Test the PRV: Perform a manual test of the PRV by lifting the lever or handle briefly to allow a small amount of water to discharge through the discharge pipe. This action should relieve some pressure and verify that the PRV is functional. If no water discharges or if the PRV fails to open, it may be stuck or faulty.

Clean the PRV: If the PRV appears to be stuck or obstructed, try cleaning it by gently tapping the valve with a wrench or mallet to dislodge any debris. You can also use compressed air to blow out any obstructions from the valve.

Replace the PRV: If cleaning the PRV does not resolve the issue, or if the PRV is damaged, it will need to be replaced. To replace the PRV, turn off the water supply to the water heater, drain some water from the tank to relieve pressure, disconnect the discharge pipe, and remove the old PRV. Install the new PRV according to the manufacturer's instructions, reconnect the discharge pipe, and turn the water supply back on.

Test the Repair: Once the new PRV is installed, test it again by lifting the lever or handle to ensure it opens and relieves pressure properly. Monitor the discharge pipe for any leaks or abnormal behavior.

Monitor Performance: Keep an eye on the water heater and PRV performance over the next few days to ensure that the issue has been resolved and that the PRV is functioning correctly.

EXHAUST VENTING ISSUES

Common issues with exhaust venting in a gas water heater can pose serious safety risks and affect the performance of the appliance. Some common issues include:

Improper Vent Installation: If the exhaust vent is not installed correctly, it can lead to improper venting of combustion gases. This may result in backdrafting, where combustion gases flow back into the living space instead of being expelled outdoors.

Vent Blockages: Blockages in the exhaust vent, such as debris, bird's nests, or ice buildup, can obstruct the flow of combustion gases. This can cause the water heater to malfunction or shut down due to inadequate venting.

Corrosion or Deterioration: Over time, the exhaust vent pipes can corrode or deteriorate, especially if they are exposed to moisture or harsh weather conditions. Corrosion can lead to holes or cracks in the vent pipes, causing gas leaks or improper venting.

Vent Size or Length: Incorrectly sized or excessively long vent pipes can lead to poor draft and inadequate venting of combustion gases. This can result in inefficient operation of the water heater and potential safety hazards.

Venting into Chimneys or Ducts: Venting a gas water heater into a chimney or duct designed for another appliance, such as a furnace, can lead to improper venting and backdrafting. Each gas appliance should have its own dedicated venting system.

Insufficient Combustion Air: Gas water heaters require adequate combustion air for proper operation. If there is not enough air supply for combustion, it can lead to incomplete combustion and the production of carbon monoxide. Ensure that combustion air vents are clear and unobstructed.

Vent Slope: The exhaust vent should be sloped upward away from the water heater to ensure proper draft and drainage of condensate. If the vent is not properly sloped, condensate can accumulate and obstruct the venting path.

Vent Termination Location: The termination point of the exhaust vent should be located in a safe and appropriate location outdoors. Improper termination locations, such as near windows, doors, or air intakes, can lead to backdrafting and safety hazards.

To address issues with exhaust venting in a gas water heater:

Inspect Regularly: Regularly inspect the exhaust vent system for signs of blockages, corrosion, or damage. Check for proper venting during operation, including the absence of backdrafting.

Clear Blockages: Remove any debris, nests, or ice buildup from the exhaust vent pipes to ensure unobstructed flow of combustion gases.

Repair or Replace: Repair or replace any damaged or corroded vent pipes to maintain proper venting and prevent gas leaks.

Ensure Proper Installation: Ensure that the exhaust vent system is installed correctly according to manufacturer guidelines and local building codes. Proper installation includes appropriate vent sizing, slope, and termination location.

Monitor Combustion Air: Ensure that there is adequate combustion air supply for the water heater by keeping combustion air vents clear and unobstructed.

CORROSION AND RUST

Corrosion and rust in a gas water heater can lead to various problems, including leaks, decreased efficiency, and premature failure of the appliance. Here are some common issues associated with corrosion and rust in gas water heaters:

Tank Leakage: Corrosion of the inner lining or tank wall can lead to pinhole leaks or cracks in the water heater tank, resulting in water leakage. This can cause water damage to the surrounding area and require immediate repair or replacement of the water heater.

Reduced Efficiency: Corrosion and rust buildup inside the water heater tank can insulate the heating elements, reducing heat transfer efficiency. This can result in longer heating times, increased energy consumption, and higher utility bills.

Discolored Water: Corrosion and rust particles can contaminate the water supply, causing the water to appear discolored or rusty. This can affect the quality and taste of the water and may indicate internal corrosion issues within the water heater.

Foul Odors: Corrosion and rust inside the water heater tank can promote the growth of bacteria and sediment, leading to foul-smelling water. This can result in unpleasant odors when using hot water and may require flushing or cleaning of the tank to resolve.

Sediment Accumulation: Corrosion and rust can contribute to sediment buildup inside the water heater tank, reducing the available storage capacity and interfering with heating performance. Excessive sediment accumulation can also lead to overheating and premature failure of the heating elements.

Tank Failure: Severe corrosion and rusting of the water heater tank can weaken its structural integrity, leading to catastrophic tank failure and flooding. This can pose a significant safety hazard and may cause extensive property damage if not addressed promptly.

To prevent or address corrosion and rust issues in a gas water heater:

Regular Maintenance: Perform annual maintenance tasks, such as flushing the water heater tank to remove sediment and inspecting for signs of corrosion or rust. Regular maintenance can help identify and address issues before they escalate.

Anode Rod Inspection: Inspect the sacrificial anode rod inside the water heater tank for signs of corrosion or depletion. Replace the anode rod if it shows significant corrosion, as it helps protect the tank from internal rusting.

Water Quality Testing: Test the water quality periodically to identify any corrosive elements or contaminants that may contribute to corrosion and rusting. Consider installing a water softener or filtration system if necessary to improve water quality.

Proper Ventilation: Ensure that the water heater is properly ventilated to prevent moisture buildup and corrosion of internal components. Adequate ventilation helps maintain a dry environment inside the water heater tank.

Temperature Regulation: Avoid setting the water heater temperature too high, as excessive heat can accelerate corrosion and rusting of internal components. Set the temperature to a moderate level to minimize corrosion risks.

ANODE ROD DETERIORATION

The sacrificial anode rod in a gas water heater plays a crucial role in protecting the tank from corrosion by attracting corrosive elements in the water. Over time, the anode rod deteriorates due to its sacrificial nature, leading to various issues. Here are common problems associated with anode rod deterioration in a gas water heater:

Tank Corrosion: As the sacrificial anode rod deteriorates, the tank becomes more vulnerable to corrosion. Corrosion can lead to pinhole leaks, cracks, or even catastrophic failure of the tank, resulting in water damage and the need for replacement.

Rusty Water: Deterioration of the anode rod can cause rust particles to accumulate in the water heater tank. These particles can discolor the hot water, making it appear rusty or discolored. Rusty water may also have an unpleasant taste and odor.

Reduced Anode Efficiency: As the anode rod deteriorates, its ability to protect the tank diminishes. This can lead to accelerated corrosion of internal components, such as the tank walls or heating elements, resulting in reduced efficiency and premature failure of the water heater.

Sediment Accumulation: Deteriorating anode rods can contribute to sediment buildup inside the water heater tank. Sediment accumulation reduces the available storage capacity, interferes with heating performance, and may lead to overheating or scaling issues.

Odor Problems: Rust and sediment accumulation caused by deteriorating anode rods can promote the growth of bacteria and microorganisms in the water heater tank. This can result in foul-smelling water and unpleasant odors when using hot water.

Anode Rod Stuck in Tank: In some cases, the anode rod may become corroded and stuck inside the tank, making it difficult to remove for replacement. This can require additional effort and may necessitate professional assistance to extract the deteriorated anode rod.

To address issues related to anode rod deterioration in a gas water heater:

Regular Inspection: Inspect the sacrificial anode rod annually for signs of deterioration, corrosion, or depletion. Check for visible signs of rust or pitting on the rod surface, as well as any significant reduction in length.

Replace Anode Rod: Replace the sacrificial anode rod if it shows signs of significant deterioration or corrosion. Anode rods are typically designed to be replaced every three to five years, depending on water quality and usage conditions.

Select Proper Anode Material: Choose an anode rod made from a suitable material for your water quality and conditions. Common materials include magnesium, aluminum, and aluminum-zinc alloy. Consult the water heater manufacturer or a qualified technician for guidance on selecting the appropriate anode rod.

Flush Tank Regularly: Perform regular maintenance tasks, such as flushing the water heater tank, to remove sediment buildup and maintain optimal performance. Flushing helps prolong the lifespan of the anode rod and prevents corrosion issues.

Monitor Water Quality: Test the water quality periodically to identify any corrosive elements or contaminants that may contribute to anode rod deterioration. Address water quality issues to minimize corrosion risks and extend the lifespan of internal components.

Garbage Disposal

Several common issues can lead to clogs in a garbage disposal.

Putting the Wrong Foods Down the Disposal: Certain foods are more likely to cause clogs than others. Avoid putting hard items like bones, fruit pits, or fibrous materials like celery, corn husks, or potato peels down the disposal, as these can get tangled in the blades or create blockages.

Insufficient Water Flow: Running the disposal without enough water can contribute to clogs. Adequate water flow helps flush food waste through the disposal and prevents particles from getting stuck in the drainpipe.

Overloading the Disposal: Overloading the disposal with too much food waste at once can overwhelm the unit and lead to clogs. It's best to feed food waste into the disposal gradually and in small amounts to prevent clogs.

Lack of Maintenance: Neglecting regular maintenance, such as cleaning the disposal or removing trapped food particles, can lead to clogs over time. Food debris can accumulate in the disposal or drainpipe, increasing the risk of blockages.

Dull Blades: Dull blades in the disposal can struggle to effectively grind food waste, leading to larger food particles that are more likely to cause clogs. Regularly sharpening or replacing the blades can help prevent this issue.

Foreign Objects: Non-food items like utensils, bottle caps, or small kitchen tools accidentally dropped into the disposal can cause clogs or damage to the unit's components.

Improper Installation: Incorrect installation of the disposal or drain piping can create bends or obstructions that impede the flow of food waste, leading to clogs.

Repairing a clogged garbage disposal typically involves several steps. Here's a guide to clearing a clog in your garbage disposal:

Turn Off the Power: Before attempting any repairs, ensure that the power to the garbage disposal is turned off to prevent accidental injury. You can usually do this by unplugging the disposal from the power outlet or turning off the circuit breaker that supplies power to the unit.

Identify the Source of the Clog: Determine where the clog is located. If the clog is within the disposal unit itself, you may hear a humming sound when you turn it on. If the clog is in the drainpipe, water may back up into the sink when you run the disposal.

Clear the Disposal Unit: If the clog is within the disposal unit, use a flashlight to inspect the inside of the disposal chamber. Look for any visible blockages or objects causing the clog, such as food waste or foreign objects. Use tongs or pliers to remove any visible debris from the disposal chamber.

Use a Plunger: If the clog persists, try using a sink plunger to clear it. Fill the sink with enough water to cover the rubber head of the plunger, place the plunger over the drain opening, and plunge vigorously several times to dislodge the clog. Repeat as needed until the water begins to drain freely.

Reset the Disposal: If the disposal became overloaded and shut off due to a clog, it may have tripped a reset button or overload protector. Locate the reset button on the bottom or side of the disposal unit and press it to reset the motor. Wait a few minutes before attempting to use the disposal again.

Use a Hex Key: Some garbage disposals come with a hex key or disposal wrench that can be inserted into the bottom of the disposal unit to manually rotate the impeller and free a jammed or stuck object. Insert the hex key into the center shaft at the bottom of the disposal and turn it clockwise and counterclockwise several times to dislodge the clog.

Run Cold Water: After clearing the clog, run cold water through the disposal for a few minutes to flush out any remaining debris and ensure proper drainage.

Test the Disposal: Once you've cleared the clog, restore power to the garbage disposal and test it to ensure that it's functioning properly. Run the disposal with cold water for a few seconds to confirm that it's draining as expected.

JAMMING

Several common issues can cause a garbage disposal to jam.

Overloading: Putting too much food waste into the disposal at once can overwhelm the unit and cause it to jam. Overloading can prevent the blades from spinning freely and grinding the food waste effectively.

Hard or Tough Foods: Hard or tough food items like bones, fruit pits, seeds, or coffee grounds can jam the disposal's blades or impeller. These items are difficult to grind and can damage the disposal's components.

Fibrous Foods: Fibrous foods like celery, corn husks, onion skins, or potato peels can wrap around the disposal's blades or impeller, causing them to become stuck and leading to a jam.

Foreign Objects: Accidentally dropping non-food items like utensils, bottle caps, or small kitchen tools into the disposal can jam the blades or impeller and prevent the unit from functioning properly.

Insufficient Water Flow: Running the disposal without enough water can contribute to jams by preventing food waste from flowing freely through the unit. Adequate water flow helps lubricate the blades and flush food particles through the drainpipe.

Dull Blades: Over time, the blades of the disposal can become dull, reducing their effectiveness in grinding food waste and increasing the risk of jams.

Lack of Maintenance: Neglecting regular maintenance, such as cleaning the disposal or removing trapped food particles, can lead to jams over time. Food debris can accumulate in the disposal or drainpipe, increasing the risk of blockages.

Repairing a jammed garbage disposal involves several steps. Here's a guide to clearing a jam in your garbage disposal:

Turn Off the Power: Before attempting any repairs, ensure that the power to the garbage disposal is turned off to prevent accidental injury. You can usually do this by unplugging the disposal from the power outlet or turning off the circuit breaker that supplies power to the unit.

Identify the Jam: Determine the location and cause of the jam. Use a flashlight to inspect the disposal chamber and look for any visible blockages or objects causing the jam, such as food waste or foreign objects.

Use a Hex Key: Many garbage disposals come with a hex key or disposal wrench that can be inserted into the bottom of the disposal unit to manually rotate the impeller and free a jammed or stuck object. Insert the hex key into the center shaft at the bottom of the disposal and turn it clockwise and counterclockwise several times to dislodge the jam.

Clear the Blockage: If you can identify and reach the source of the jam, use tongs, pliers, or a long-handled spoon to remove any visible debris or objects from the disposal chamber. Be careful not to use your hands, and never reach into the disposal while it's powered on.

Reset the Disposal: If the disposal became overloaded and shut off due to the jam, it may have tripped a reset button or overload protector. Locate the reset button on the bottom or side of the disposal unit and press it to reset the motor. Wait a few minutes before attempting to use the disposal again.

Run Cold Water: After clearing the jam, run cold water through the disposal for a few minutes to flush out any remaining debris and ensure proper drainage.

Test the Disposal: Once you've cleared the jam, restore power to the garbage disposal and test it to ensure that it's functioning properly. Run the disposal with cold water for a few seconds to confirm that it's draining as expected.

FOUL ODORS

Foul odors in a garbage disposal are typically caused by the accumulation of food debris, bacteria, and organic matter. Here are common issues that can contribute to foul odors in a garbage disposal:

Food Residue Buildup: Over time, small food particles can accumulate on the blades, walls, and other surfaces inside the disposal. When these particles decompose, they can produce unpleasant odors.

Bacterial Growth: The warm and damp environment inside the disposal provides an ideal breeding ground for bacteria. Bacteria breaking down food particles can release gases that contribute to foul odors.

Neglected Cleaning: Lack of regular cleaning and maintenance allows food debris to build up in the disposal, leading to the growth of bacteria and the development of odors.

Fibrous Foods: Certain fibrous foods, such as onion skins, celery, or citrus peels, can get tangled in the disposal blades and create a breeding ground for bacteria. As these materials break down, they contribute to foul smells.

Lack of Water Flow: Running the disposal without sufficient water flow can result in incomplete flushing of food particles, allowing them to stick to the disposal surfaces and contribute to odor problems.

To address foul odors in a garbage disposal:

Run Cold Water: Always run cold water while operating the disposal and for a few seconds afterward. Cold water helps flush food particles through the disposal and into the drainpipe, preventing them from accumulating and causing odors.

Clean the Disposal Regularly: Clean the disposal regularly to remove food debris and prevent bacterial growth. You can clean the disposal by grinding ice cubes, citrus peels, or a mixture of baking soda and vinegar. This helps dislodge debris and eliminate odors.

Use Disposal Cleaner: Consider using a commercial garbage disposal cleaner to break down organic matter and eliminate odors. Follow the manufacturer's instructions for proper usage.

Avoid Putting Certain Foods Down the Disposal: Avoid putting fibrous or strongly scented foods down the disposal, as these can contribute to odor problems. Instead, dispose of such items in the trash.

Run the Disposal Regularly: Even if you don't have food waste to dispose of, run the disposal regularly to prevent stagnation and keep the blades and chamber clean.

Deodorize with Citrus: Grind citrus peels, such as lemon or orange, in the disposal to provide a fresh and natural deodorizing effect.

Check for Foreign Objects: Occasionally inspect the disposal for any foreign objects or items that might have accidentally fallen in and contribute to odors.

LEAKING

Common issues that can cause a garbage disposal to leak include:

Worn or Damaged Seals: Over time, the seals between the various components of the garbage disposal can wear out or become damaged, leading to leaks. This can include the seal between the sink flange and the disposal unit, as well as the seals around the disposal's motor housing.

Loose Connections: Loose connections between the disposal unit and the drainpipe, dishwasher hose, or other plumbing components can allow water to escape and cause leaks. This can occur if the disposal wasn't properly installed or if the connections have become loose over time.

Cracked or Damaged Housing: The housing of the garbage disposal unit itself can develop cracks or other damage, especially if the disposal has been subjected to impact or stress. Cracks in the housing can allow water to seep out and cause leaks.

Faulty Dishwasher Connection: If your garbage disposal is connected to a dishwasher, a leak can occur at the connection point if the hose is damaged, improperly installed, or not securely attached.

Worn Out Gaskets or O-Rings: Gaskets and O-rings inside the disposal unit can deteriorate over time, leading to leaks. These components help create a watertight seal between the various parts of the disposal, and if they become worn or damaged, leaks can occur.

Cracked or Damaged Drainpipe: Leaks can also originate from the drainpipe connected to the disposal. If the pipe is cracked, corroded, or otherwise damaged, water can leak out and cause problems.

Excessive Pressure or Backups: If the garbage disposal becomes clogged or experiences a backup in the drain line, it can create excessive pressure inside the unit, leading to leaks. This can occur if the disposal is overloaded with food waste or if there are blockages in the drain line.

To address a leaking garbage disposal:

Inspect for Visible Leaks: Check around the disposal unit, under the sink, and along the drainpipe for any signs of water accumulation or moisture that may indicate a leak.

Tighten Connections: Ensure that all connections between the disposal unit, drainpipe, dishwasher hose, and other plumbing components are tight and secure. Use a wrench or pliers to tighten any loose connections.

Replace Seals or Gaskets: If you identify worn or damaged seals, gaskets, O-rings, or other components, replace them as needed to restore the watertight seal.

Repair Cracks or Damage: If the disposal unit's housing is cracked or damaged, consider replacing the unit to prevent further leaks. If the drainpipe is damaged, repair or replace it as necessary.

Check the Dishwasher Connection: Inspect the connection between the garbage disposal and the dishwasher for any signs of damage or leaks. Replace the dishwasher hose or connection if necessary.

Clear Clogs and Backups: If the disposal is experiencing clogs or backups, clear them to relieve pressure and prevent leaks. Use a plunger or plumbing snake to clear any blockages in the drain line.

ELECTRICAL ISSUES

Common electrical issues with a garbage disposal may include:

Power Supply Problems: Issues with the power supply, such as a tripped circuit breaker, a blown fuse, or a loose electrical connection, can prevent the garbage disposal from receiving power and functioning properly.

Faulty Wiring: Wiring problems within the garbage disposal unit, such as frayed or damaged wires, loose connections, or short circuits, can lead to electrical malfunctions and pose safety hazards.

Defective Switch: A malfunctioning or defective wall switch that controls the garbage disposal can prevent it from turning on or off properly. This may be due to a faulty switch mechanism, loose wiring connections, or internal switch issues.

Motor Problems: Electrical issues with the motor, such as overheating, wiring faults, or internal component failures, can cause the garbage disposal to malfunction or stop working altogether.

Ground Fault Circuit Interrupter (GFCI) Tripping: Garbage disposals installed in areas where GFCI protection is required may trip the GFCI outlet if there is a ground fault or electrical imbalance. This can be caused by water intrusion, damaged wiring, or internal faults within the disposal unit.

Electrical Shock Hazard: Faulty wiring, damaged insulation, or improper installation can create electrical shock hazards, posing risks to anyone working on or around the garbage disposal.

Intermittent Operation: Intermittent operation of the garbage disposal, where it turns on and off unexpectedly or inconsistently, may indicate underlying electrical issues such as loose connections, faulty switches, or motor problems.

To address electrical issues with a garbage disposal:

Check Power Supply: Verify that the garbage disposal is receiving power by checking the circuit breaker or fuse box for tripped breakers or blown fuses. Reset any tripped breakers or replace blown fuses as needed.

Inspect Wiring: Inspect the wiring connections inside the garbage disposal unit and at the electrical junction box for any signs of damage, corrosion, or loose connections. Tighten loose connections and repair or replace damaged wiring as necessary.

Test the Switch: Test the wall switch that controls the garbage disposal to ensure it is functioning properly. If the switch is faulty, replace it with a new one.

Reset the Disposal: If the garbage disposal has a reset button, press it to reset the unit and see if it resolves the electrical issue.

Check GFCI Protection: If the garbage disposal is connected to a GFCI outlet, ensure that the outlet is not tripped and that it is providing power to the disposal. Test the GFCI outlet and reset it if necessary.

Inspect Motor: If the garbage disposal motor is not running or is making unusual noises, it may indicate internal electrical issues. Consult the manufacturer's instructions for troubleshooting or consider contacting a qualified appliance repair technician for further inspection and repair.

Safety Precautions: When working on or around the garbage disposal, always turn off power to the unit at the circuit breaker or fuse box to prevent electrical shock hazards.

Electrical Sockets

LOOSE CONNECTIONS

Common issues associated with a loose connection in an electrical socket include:

Intermittent Power Supply: A loose connection can cause intermittent power supply to devices plugged into the socket. Devices may turn on and off unexpectedly, or their operation may be disrupted.

Arcing and Sparking: Loose connections can create arcing and sparking within the socket, which can lead to overheating, melting, or ignition of nearby flammable materials. This poses a fire hazard and can cause damage to the socket and connected devices.

Heat Buildup: Loose connections generate resistance in the electrical circuit, leading to heat buildup at the point of connection. This can cause the socket, wiring, or surrounding materials to become hot to the touch and may result in thermal damage or fire.

High Electrical Resistance: Loose connections increase electrical resistance in the circuit, which can affect the performance of connected devices and lead to inefficiencies in power distribution.

Damage to Wiring: Prolonged exposure to heat and arcing caused by loose connections can damage the wiring within the socket and the electrical circuit. This can compromise the integrity of the wiring and increase the risk of electrical faults and hazards.

Safety Hazard: Loose connections pose a significant safety hazard, as they increase the risk of electrical shock, fire, and property damage. They should be addressed promptly to prevent accidents and ensure the safe operation of electrical systems.

To address a loose connection in an electrical socket:

Turn Off Power: Before attempting any repairs, turn off power to the affected electrical circuit at the circuit breaker panel to prevent the risk of electrical shock or fire.

Inspect the Socket: Carefully inspect the electrical socket for signs of damage, burning, or discoloration that may indicate a loose connection. Look for any visible gaps or looseness in the wiring connections.

Tighten Connections: If you identify a loose connection, use a screwdriver to tighten the terminal screws or wire connections inside the socket. Ensure that the wires are securely fastened and that there is no play or movement in the connections.

Replace Damaged Components: If the socket or wiring is damaged beyond repair, replace the damaged components with new ones. This may involve replacing the socket, rewiring the connections, or repairing any damaged wiring.

Test the Socket: After making repairs, test the socket to ensure that it is functioning properly and that there are no signs of overheating, sparking, or arcing. Plug in a device and verify that it receives power without any issues.

OVERLOADING

Overloading an electrical socket occurs when the total electrical load connected to the socket exceeds its maximum capacity. Common issues associated with overloading an electrical socket include:

Tripped Circuit Breaker: An overloaded socket can trip the circuit breaker or blow a fuse in the electrical panel, cutting off power to the affected circuit. This is a safety feature designed to prevent overheating and electrical fires.

Overheating: Excessive electrical current flowing through an overloaded socket can cause the wiring, outlet, or connected devices to overheat. Overheating can lead to damage, melting, or ignition of nearby materials, posing a fire hazard.

Dimming Lights: If the electrical circuit is overloaded, it may cause lights connected to the same circuit to dim or flicker as the voltage drops due to increased electrical demand.

Electrical Shock Hazard: Overloading a socket can increase the risk of electrical shock, especially if damaged or improperly insulated wiring is involved. Contact with live electrical components can result in serious injury or death.

Damage to Devices: Overloading a socket with more devices than it can safely handle can damage the devices themselves. Excessive electrical current can cause overheating, short circuits, or other electrical faults in connected devices.

Reduced Efficiency: Overloading a socket can lead to reduced efficiency and performance of connected devices, as they may not receive adequate power to operate properly. This can lead to malfunctions, data loss, or damage to electronic equipment.

Repairing an overloaded electrical socket involves several steps to ensure safety and proper functionality. Here's a guide to repairing an overloaded electrical socket:

Turn Off Power: Before attempting any repairs, turn off the power to the affected electrical circuit at the circuit breaker panel to prevent the risk of electrical shock or fire.

Identify Overloaded Socket: Identify the socket that has been overloaded. Look for signs such as a tripped circuit breaker, flickering lights, or charred marks around the outlet.

Unplug Devices: Disconnect all devices from the overloaded socket to reduce the electrical load. This includes unplugging any appliances, chargers, or other devices connected to the socket.

Inspect Wiring and Outlet: Carefully inspect the wiring and outlet for signs of damage, overheating, or loose connections. Look for melted insulation, blackened or charred areas, or loose terminal screws.

Tighten Connections: If the wiring connections are loose, use a screwdriver to tighten the terminal screws on the outlet. Ensure that the wires are securely fastened and that there is no play or movement in the connections.

Check for Damage: If the outlet or wiring shows signs of damage beyond repair, such as melted insulation or charred components, the outlet should be replaced. Consult a qualified electrician for assistance with replacing the outlet and inspecting the wiring.

Reset Circuit Breaker: If the circuit breaker has tripped due to overloading, reset it by flipping the breaker switch to the "off" position and then back to the "on" position. If the breaker continues to trip, it may indicate a persistent overload or wiring issue that requires further investigation by a professional.

Distribute Electrical Load: Avoid overloading the repaired socket in the future by distributing electrical loads across multiple sockets and circuits. Use power strips or surge protectors with built-in overload protection to safely connect multiple devices.

Test Socket: After making repairs and redistributing electrical loads, test the socket to ensure that it is functioning properly. Plug in a device and verify that it receives power without any issues.

Monitor for Further Problems: Monitor the repaired socket and electrical circuit for any signs of overheating, sparking, or abnormal operation.

FAULTY WIRING

Faulty wiring in electrical sockets can lead to various issues, including:

Intermittent Power Supply: Faulty wiring can cause intermittent power supply to devices plugged into the socket. Devices may turn on and off unexpectedly, or their operation may be disrupted.

Overheating: Poorly installed or damaged wiring can lead to overheating of the electrical socket and surrounding components. Overheating can cause melting, burning, or ignition of nearby materials, posing a fire hazard.

Electrical Shock Hazard: Faulty wiring increases the risk of electrical shock, especially if damaged or improperly insulated wires are involved. Contact with live electrical components can result in serious injury or death.

High Electrical Resistance: Faulty wiring increases electrical resistance in the circuit, leading to inefficiencies in power distribution and reduced performance of connected devices.

Tripped Circuit Breakers or Blown Fuses: Faulty wiring can cause circuit breakers to trip or fuses to blow, cutting off power to the affected circuit. This is a safety mechanism designed to prevent overheating and electrical fires.

Short Circuits: Faulty wiring can create short circuits, where an electrical current bypasses its intended path and flows directly from the hot wire to the neutral or ground wire. Short circuits can cause sparks, arcing, and electrical fires.

Electrical Code Violations: Faulty wiring that does not meet electrical code requirements can result in violations and may need to be corrected to ensure compliance and safety.

Dimming or Flickering Lights: Faulty wiring can cause lights connected to the same circuit as the socket to dim or flicker as the voltage drops due to increased electrical resistance.

To address issues with faulty wiring in electrical sockets:

Turn Off Power: Before attempting any repairs, turn off the power to the affected electrical circuit at the circuit breaker panel to prevent the risk of electrical shock or fire.

Inspect Wiring: Carefully inspect the wiring inside the electrical socket and the wiring leading to it for signs of damage, wear, or improper installation. Look for frayed insulation, loose connections, or exposed wires.

Repair or Replace Wiring: If you identify faulty wiring, repair or replace it as necessary. This may involve splicing wires, replacing damaged sections of wiring, or rerouting wiring to ensure safe and proper installation.

Tighten Connections: Ensure that all wiring connections are tight and secure. Use wire nuts, terminal screws, or other appropriate connectors to secure wires together and prevent loosening or separation.

Replace Electrical Components: If the electrical socket or other components are damaged beyond repair, replace them with new ones. This may involve replacing the socket, junction box, or wiring as necessary.

TRIPPED BREAKER

Several common causes can lead to a tripped circuit breaker from an electrical socket:

Overload: One of the most common reasons for a circuit breaker to trip is when the electrical load exceeds the circuit's capacity. This can happen if too many devices are plugged into the same circuit or if a high-power device is used simultaneously with other devices on the same circuit.

Short Circuit: A short circuit occurs when a hot wire comes into direct contact with a neutral or ground wire, bypassing the resistance of the load. This can happen due to damaged insulation, loose connections, or faulty wiring. Short circuits cause a sudden surge of current, triggering the circuit breaker to trip to prevent overheating and a fire.

Ground Fault: A ground fault occurs when a hot wire comes into contact with a ground wire or a grounded metal object, creating a path for electrical current to flow to the ground. Ground faults can occur due to damaged appliances, wet conditions, or faulty wiring. Ground fault circuit interrupters (GFCIs) are designed to detect ground faults and trip the circuit breaker to prevent electrical shock.

Arc Fault: An arc fault occurs when electricity jumps between two or more conductors, creating an arc or spark. Arc faults can happen due to damaged wiring, loose connections, or deteriorated insulation. Arc fault circuit interrupters (AFCIs) are designed to detect arc faults and trip the circuit breaker to prevent electrical fires.

Weak Circuit Breaker: Over time, circuit breakers can become worn out or weakened, reducing their ability to handle electrical loads and increasing the likelihood of tripping. In such cases, replacing the circuit breaker with a new one may be necessary to ensure proper protection.

Temperature Changes: Extreme temperature changes, such as those experienced during heatwaves or cold snaps, can affect the electrical resistance of wires and components, leading to fluctuations in current flow and tripping of circuit breakers.

Electrical System Issues: Problems with the electrical system, such as loose connections in the main electrical panel, damaged wiring, or faulty components, can also cause circuit breakers to trip. These issues may require professional inspection and repair to identify and address.

To repair a tripped circuit breaker from an electrical socket, follow these steps:

Identify the Cause: Determine the cause of the circuit breaker trip by assessing the circumstances leading up to the event. Consider factors such as the devices plugged into the circuit, recent changes or additions to the electrical system, and any visible signs of damage or wear.

Turn Off Power: Before attempting any repairs, turn off the power to the affected circuit at the circuit breaker panel to prevent the risk of electrical shock or injury. Locate the circuit breaker corresponding to the tripped circuit and switch it to the "off" position.

Inspect Wiring and Outlets: Carefully inspect the electrical wiring and outlets connected to the circuit for signs of damage, wear, or loose connections. Look for frayed insulation, exposed wires, burn marks, or other abnormalities that may indicate a problem.

Address Overload: If the circuit was tripped due to an overload, unplug some devices from the affected circuit to reduce the electrical load. Consider redistributing devices across multiple circuits or using power strips with built-in overload protection to prevent future overloads.

Reset Circuit Breaker: After addressing the cause of the circuit breaker trip, reset the circuit breaker by firmly switching it back to the "on" position. If the circuit breaker trips immediately upon resetting, it may indicate a persistent issue that requires further investigation.

Test Circuit: Test the circuit to ensure that power is restored and that devices plugged into the electrical sockets are functioning properly. Plug in a device and verify that it receives power without any issues.

Monitor for Further Problems: Monitor the circuit for any signs of recurring issues, such as repeated tripping of the circuit breaker or abnormal operation of devices connected to the circuit. If problems persist, it may indicate an underlying electrical problem that requires professional inspection and repair.

DAMAGED SOCKET

Common issues of a damaged electrical socket may include:

Loose Connections: Damage to the internal wiring or terminal connections of the socket can result in loose connections, leading to intermittent power supply or complete loss of power to devices plugged into the socket.

Burn Marks or Scorching: Overheating due to loose connections, overload, or faulty wiring can cause burn marks or scorching around the socket, indicating potential fire hazards and the need for immediate attention.

Cracks or Breaks: Physical damage to the socket, such as cracks or breaks in the housing or faceplate, can compromise its structural integrity and pose safety risks. Cracks may allow moisture or debris to enter the socket, increasing the risk of electrical faults.

Exposed Wiring: Damage to the insulation or casing of the wiring within the socket can expose live wires, increasing the risk of electrical shock and posing a safety hazard to anyone using the socket.

Intermittent Power Supply: A damaged socket may provide intermittent power supply to connected devices, resulting in flickering lights, device malfunctions, or unexpected power outages.

Electrical Arcing: Faulty wiring or loose connections within the socket can cause electrical arcing, where electricity jumps between conductors, producing sparks or arcs. Electrical arcing can damage the socket and surrounding materials and pose fire hazards.

Inoperable Switches or Outlets: Damage to the internal mechanisms of switches or outlets within the socket can render them inoperable, preventing devices from receiving power or being controlled.

Corrosion or Rust: Exposure to moisture or environmental factors can cause corrosion or rust on the metal components of the socket, compromising its conductivity and longevity.

GFCI or AFCI Failure: Damage to the internal components of ground fault circuit interrupters (GFCIs) or arc fault circuit interrupters (AFCIs) integrated into the socket can result in their failure to detect ground faults or arc faults, reducing their protective capabilities.

Overloading: Continual overloading of the socket with high-power devices or appliances can lead to damage and premature wear, reducing the socket's lifespan and increasing the risk of electrical faults.

To address issues with a damaged electrical socket:

Turn Off Power: Before attempting any repairs, turn off power to the affected circuit at the circuit breaker panel to prevent the risk of electrical shock or fire.

Inspect Socket: Carefully inspect the socket for signs of damage, including burn marks, cracks, exposed wiring, or corrosion. If the damage is severe or poses immediate safety risks, refrain from using the socket and consult a qualified electrician for repair or replacement.

Replace Damaged Components: If the socket or internal components are damaged beyond repair, replace them with new ones. This may involve replacing the socket, faceplate, wiring, or switches as necessary.

Address Wiring Issues: If the damage extends to the wiring within the socket or the electrical system, consult a qualified electrician to inspect and repair the wiring to ensure safe and proper operation.

Test Functionality: After making repairs or replacements, test the socket to ensure that it is functioning properly and safely. Plug in a device and verify that it receives power without any issues.

GROUND FAULT ISSUES

Common issues of ground faults in electrical sockets include:

Electrical Shocks: Ground faults can create a pathway for electrical current to flow through unintended routes, such as through a person's body. This can result in electrical shocks, which can range from mild to severe depending on the current flow and duration.

Tripped GFCI: Ground fault circuit interrupters (GFCIs) are designed to detect ground faults and quickly interrupt the circuit to prevent electrical shocks and fires. If a ground fault occurs, the GFCI will trip, cutting off power to the affected outlet and potentially other outlets on the same circuit.

Potential Fire Hazard: Ground faults can cause overheating of electrical wiring and components, posing a fire hazard. The arcing or sparking associated with ground faults can ignite nearby materials, leading to electrical fires.

Interference with Electronic Equipment: Ground faults can cause electrical interference or noise in sensitive electronic equipment, leading to malfunctions, data loss, or damage to electronic devices.

Damage to Electrical Components: Continual ground faults can damage electrical components, including wiring, outlets, switches, and connected devices. The repeated interruption of power and the associated arcing can degrade the integrity of these components over time.

Corrosion and Deterioration: Ground faults can lead to corrosion or deterioration of electrical wiring and components, especially if moisture or environmental factors are involved. This can further exacerbate the risk of electrical faults and safety hazards.

Diminished Safety Protections: In residential and commercial settings, ground faults can compromise the effectiveness of safety mechanisms such as GFCIs and AFCIs. Faulty or malfunctioning GFCIs may fail to detect ground faults, leaving occupants vulnerable to electrical hazards.

Electrical Code Violations: Ground faults can result in electrical code violations if not addressed promptly. Failure to comply with electrical codes and regulations can lead to fines, penalties, and increased liability in the event of accidents or injuries.

To address ground faults in electrical sockets:

Test GFCIs: Regularly test GFCI outlets using the test button to ensure they are functioning properly. If a GFCI trips repeatedly or fails to reset, it may indicate a ground fault that requires further investigation.

Inspect Wiring and Outlets: Carefully inspect electrical wiring, outlets, and switches for signs of damage, wear, or corrosion. Look for burn marks, exposed wires, loose connections, or other abnormalities that may indicate a ground fault.

Address Wiring Issues: If ground faults are detected, consult a qualified electrician to inspect and repair the wiring to ensure safe and proper operation. This may involve identifying and correcting faulty wiring, loose connections, or damaged components.

Replace Faulty Components: If electrical outlets, switches, or GFCIs are found to be faulty or damaged, replace them with new ones. Ensure that replacements are installed correctly and comply with electrical codes and regulations.

Reduce Moisture Exposure: Minimize exposure to moisture or environmental factors that can contribute to ground faults. Use weatherproof covers for outdoor outlets, address leaks or moisture intrusion, and avoid using electrical equipment in wet or damp conditions.

Educate Occupants: Educate occupants about the importance of electrical safety and how to recognize and respond to ground faults. Encourage safe practices such as testing GFCIs regularly, avoiding overloading circuits, and using electrical equipment properly.

WATER DAMAGE

Water-damaged electrical sockets can lead to various issues, including:

Electrical Shorts: Water can create a conductive path between electrical conductors, resulting in short circuits. Short circuits can cause immediate power loss, tripped circuit breakers, or even electrical fires if left unchecked.

Corrosion: Water exposure can cause corrosion of metal components within the socket, such as terminals, wiring, and connections. Corrosion can lead to poor electrical conductivity, increased resistance, and eventual failure of the socket.

Electrical Hazards: Water-damaged sockets pose a significant electrical hazard, increasing the risk of electrical shock or electrocution for anyone coming into contact with the socket or connected devices.

Malfunctioning Devices: Water damage can cause connected devices or appliances to malfunction or fail to operate properly. This can include flickering lights, malfunctioning switches, or damaged electronics.

Shortened Lifespan: Continued exposure to water can accelerate the deterioration of electrical components within the socket, leading to a shortened lifespan and eventual failure of the socket.

Mold and Mildew Growth: Water damage can create ideal conditions for mold and mildew growth within the electrical socket and surrounding area. Mold and mildew can compromise indoor air quality and pose health risks to occupants.

Fire Hazard: If water-damaged sockets are not properly addressed, they can pose a significant fire hazard. Water can cause insulation materials to degrade, increasing the risk of electrical arcing, sparking, and ignition of nearby materials.

Electrical Code Violations: Water-damaged sockets may violate electrical codes and regulations, which require electrical installations to be safe and free from hazards. Failure to address water damage promptly can result in fines, penalties, or legal liabilities.

To address issues from water-damaged electrical sockets:

Turn Off Power: Before attempting any repairs, turn off power to the affected circuit at the circuit breaker panel to prevent the risk of electrical shock or fire.

Inspect for Water Damage: Carefully inspect the electrical socket and surrounding area for signs of water damage, such as moisture, discoloration, rust, or corrosion. If water is present, do not attempt to use the socket until it has been properly dried and repaired.

Dry Out the Area: Thoroughly dry out the affected area using towels, fans, or dehumidifiers to remove any remaining moisture. Ensure that the area is completely dry before attempting any repairs or restoring power to the circuit.

Replace Damaged Components: Replace any damaged components within the electrical socket, including wiring, terminals, switches, or outlets. This may involve cutting back damaged wiring, replacing corroded components, or installing new sockets as necessary.

Ensure Proper Sealing: Seal any gaps or openings around the electrical socket to prevent future water intrusion. Use waterproof sealant or caulking to seal gaps and ensure a watertight seal.

Test Functionality: After making repairs, test the electrical socket to ensure that it is functioning properly and safely. Plug in a device and verify that it receives power without any issues.

Monitor for Further Problems: Monitor the repaired socket and surrounding area for any signs of recurring water damage or electrical issues. Address any new issues promptly to prevent further damage and ensure the safety of occupants.

RODENT AND PEST DAMAGE

Rodents and pests can cause various issues to electrical sockets, including:

Damage to Wiring: Rodents such as mice and rats are known to chew on electrical wiring, including wires inside electrical sockets. This can lead to exposed wires, short circuits, and electrical fires.

Short Circuits: Damage to electrical wiring by rodents can result in short circuits, where an electrical current bypasses its intended path. Short circuits can cause power outages, tripped circuit breakers, and fire hazards.

Arcing and Sparks: Rodents can cause damage to electrical components within sockets, leading to arcing or sparks. This can create fire hazards and increase the risk of electrical shocks.

Malfunctioning Devices: Damage to electrical sockets or wiring by rodents can result in malfunctioning devices or appliances connected to the affected circuit. This can include flickering lights, intermittent power supply, or failure of electronic equipment.

Electrical Hazards: Rodents nesting near electrical sockets or wiring can create electrical hazards for occupants. Nesting materials, urine, or feces can increase the risk of short circuits, electrical fires, and health hazards.

Corrosion: Rodent urine and feces can cause corrosion of metal components within electrical sockets, including terminals, wiring, and connections. Corrosion can degrade electrical conductivity and lead to poor connections and increased resistance.

Tripped Circuit Breakers: Damage to electrical wiring by rodents can cause circuit breakers to trip repeatedly, cutting off power to the affected circuit. This can disrupt power supply to connected devices and indicate underlying electrical problems.

Health Hazards: Rodents and pests nesting near electrical sockets can introduce health hazards to occupants. Rodent droppings, urine, and nesting materials can harbor pathogens, allergens, and respiratory irritants.

To address issues to electrical sockets due to rodents and pests:

Rodent Control: Implement measures to control rodents and pests in and around the property. This may include sealing entry points, removing food sources, setting traps, or using baits or repellents.

Inspect Electrical Wiring: Regularly inspect electrical wiring, outlets, and switches for signs of damage caused by rodents or pests. Look for chewed wires, nesting materials, or droppings near electrical components.

Repair Damage: Repair any damage to electrical wiring or components caused by rodents or pests. This may involve replacing damaged wires, repairing insulation, or installing protective barriers to prevent further damage.

Seal Entry Points: Seal any gaps, cracks, or openings in walls, floors, or ceilings to prevent rodents and pests from accessing electrical wiring or sockets. Use materials such as steel wool, wire mesh, or sealant to block entry points effectively.

Clean and Disinfect: Clean and disinfect areas affected by rodents or pests to remove nesting materials, droppings, and urine. Wear appropriate protective gear, such as gloves and masks, when cleaning contaminated areas.

Monitor Regularly: Monitor the property regularly for signs of rodent or pest activity, such as droppings, tracks, or chew marks. Address any new infestations promptly to prevent further damage to electrical systems.

Washing Machine

FAILURE TO DRAIN

A washing machine failing to drain properly can be frustrating, but common issues causing this problem include:

Clogged Drain Hose: The drain hose may be obstructed by lint, debris, or foreign objects, preventing water from flowing out of the machine.

Clogged Pump Filter: The pump filter, located at the front or bottom of the machine, can become clogged with lint, coins, or other debris, impeding water drainage.

Faulty Drain Pump: The drain pump may be malfunctioning or worn out, preventing it from effectively pumping water out of the machine.

Kinked or Restricted Drain Hose: The drain hose may be kinked, pinched, or restricted, hindering the flow of water and causing drainage problems.

Improper Installation of Drain Hose: If the drain hose is installed incorrectly or not positioned at the correct height, it may not be able to drain water effectively.

Clogged or Faulty Drain Valve: The drain valve, located inside the machine, may be clogged or malfunctioning, preventing water from draining properly.

Faulty Lid Switch or Door Lock: If the lid switch or door lock is faulty, the washing machine may not drain water as a safety measure to prevent operation with the lid open.

Blocked Drainpipe: The drainpipe connected to the household plumbing may be blocked or obstructed, preventing water from draining properly.

To address these issues and fix a washing machine failing to drain:

Check the Drain Hose: Inspect the drain hose for any kinks, twists, or obstructions. Remove any debris or foreign objects blocking the hose.

Clean the Pump Filter: Locate and clean the pump filter according to the manufacturer's instructions. Remove any lint, coins, or debris trapped in the filter.

Test the Drain Pump: Manually test the drain pump to ensure it is functioning properly. If the pump is making unusual noises or not running, it may need to be replaced.

Inspect the Drain Valve: Check the drain valve for any obstructions or debris. Clean or replace the drain valve if necessary.

Ensure Proper Installation of Drain Hose: Verify that the drain hose is installed correctly and positioned at the correct height. The hose should not be kinked or pinched.

Check the Lid Switch or Door Lock: Test the lid switch or door lock to ensure it is functioning properly. Replace the switch or lock if it is faulty.

Clear Blocked Drainpipe: If the drainpipe is blocked, use a plumber's snake or auger to remove the obstruction. Ensure the drainpipe is properly connected and not restricted.

LEAKS

Leaking washing machines can be a nuisance and lead to water damage in your home. Here are common issues that can cause leaks:

Worn or Damaged Door Seal: The door seal, also known as a gasket or boot seal, can degrade over time due to wear and tear, leading to leaks around the door area.

Loose or Damaged Hoses: The inlet hoses that supply water to the washing machine or the drain hose that removes water from the machine can become loose, cracked, or damaged, causing leaks.

Clogged Drain Pump or Filter: A clogged drain pump or filter can lead to water backing up and leaking out of the machine. This is often accompanied by drainage problems and error codes indicating a drain issue.

Overloading: Overloading the washing machine with too many clothes can cause excess water to spill out of the machine during the wash or spin cycle, leading to leaks.

Unbalanced Load: An unbalanced load of laundry can cause the washing machine to vibrate excessively during the spin cycle, leading to leaks from the door or other areas of the machine.

Cracked or Damaged Tub: A cracked or damaged inner or outer tub can allow water to leak out of the machine during the wash or spin cycle.

Faulty Dispenser or Detergent Drawer: A malfunctioning detergent dispenser or detergent drawer can cause water to leak out of the machine during the wash cycle.

Faulty Water Inlet Valve: A faulty water inlet valve can cause water to leak into the washing machine when it's not in use, leading to puddles of water around the machine.

Damaged or Loose Tub-to-Pump Hose: The hose that connects the tub to the pump can become damaged or come loose, causing water to leak out of the machine.

Cracked or Damaged Drain Hose: The drain hose that removes water from the machine can become cracked or damaged, causing leaks during the drain cycle.

To address leaks in a washing machine:

Inspect the Door Seal: Check the door seal for signs of wear, tears, or damage. Replace the door seal if necessary.

Tighten or Replace Hoses: Inspect the inlet hoses and drain hose for any leaks, cracks, or damage. Tighten connections or replace hoses if needed.

Clean the Drain Pump and Filter: Remove any debris or clogs from the drain pump and filter. Clean or replace the filter as needed.

Avoid Overloading: Follow the manufacturer's guidelines for loading the washing machine to prevent overloading and minimize leaks.

Balance the Load: Ensure that the load of laundry is evenly distributed inside the washing machine to prevent excessive vibration during the spin cycle.

Inspect for Tub Damage: Inspect the inner and outer tub for any cracks or damage. Replace the tub if necessary.

Check Dispenser or Detergent Drawer: Inspect the detergent dispenser or detergent drawer for any cracks or damage. Clean or replace the dispenser as needed.

Inspect Water Inlet Valve: Check the water inlet valve for leaks or damage. Replace the valve if necessary.

Check Tub-to-Pump Hose: Inspect the tub-to-pump hose for any leaks, cracks, or damage. Tighten connections or replace the hose as needed.

Inspect Drain Hose: Check the drain hose for any leaks, cracks, or damage. Replace the hose if necessary.

LOUD DURING OPERATION

A washing machine that is loud during operation can be disruptive and may indicate underlying issues. Common causes of loud washing machine operation include:

Unbalanced Load: An unbalanced load can cause the washing machine to vibrate excessively during the spin cycle, resulting in loud noises. This occurs when the laundry is not evenly distributed around the drum.

Worn Suspension Springs or Dampers: The suspension springs or dampers that support the drum may become worn or damaged over time, leading to excessive movement and noise during operation.

Loose Components: Loose or worn components within the washing machine, such as belts, pulleys, or bearings, can cause rattling or banging noises during operation.

Faulty Bearings: Worn or faulty bearings in the drum or motor can produce loud grinding or rumbling noises during the spin cycle.

Foreign Objects: Small objects, such as coins, buttons, or debris, may become trapped between the drum and the tub, causing scraping or rattling noises during operation.

Worn Drive Belt: A worn or damaged drive belt can cause squealing or squeaking noises during the wash or spin cycle.

Blocked Drain Pump or Filter: A blocked drain pump or filter can cause water to back up in the washing machine, resulting in loud noises and drainage issues.

Uneven Flooring: Uneven or unstable flooring beneath the washing machine can cause it to vibrate or shake excessively during operation, leading to increased noise levels.

To address loud washing machine operation:

Balance the Load: Ensure that the laundry is evenly distributed around the drum to prevent unbalanced loads. Avoid overloading the machine and mix different types of clothing within the load to distribute weight evenly.

Inspect Suspension Springs or Dampers: Check the suspension springs or dampers for signs of wear or damage. Replace any worn or damaged components as needed to stabilize the drum.

Tighten Loose Components: Inspect the washing machine for loose or worn components, such as belts, pulleys, or bearings. Tighten or replace any loose or worn parts to reduce noise levels.

Check for Foreign Objects: Inspect the drum and tub for any foreign objects or debris that may be causing noise during operation. Remove any objects and clean the drum and tub thoroughly.

Inspect and Replace Bearings: If the washing machine is making loud grinding or rumbling noises, the bearings may be worn or faulty. Consult a technician to inspect and replace the bearings if necessary.

Inspect Drive Belt: Check the drive belt for signs of wear or damage. Replace the belt if it appears worn or stretched.

Clean Drain Pump and Filter: Remove any debris or obstructions from the drain pump and filter to ensure proper drainage and reduce noise levels.

Stabilize the Machine: Ensure that the washing machine is level and stable on the floor. Use leveling feet to adjust the height of the machine and stabilize it on uneven flooring.

FAILURE TO SPIN

A washing machine failing to spin properly can be frustrating and may result in clothes that are not thoroughly cleaned or rinsed. Common issues causing this problem include:

Unbalanced Load: An unbalanced load of laundry can prevent the washing machine from spinning properly. This often occurs when heavy items are grouped on one side of the drum, causing imbalance during the spin cycle.

Faulty Lid Switch or Door Lock: If the lid switch or door lock is faulty, the washing machine may not engage the spin cycle as a safety measure to prevent operation with the lid open.

Worn or Broken Drive Belt: The drive belt transfers power from the motor to the drum during the spin cycle. If the belt is worn, stretched, or broken, it may prevent the drum from spinning.

Blocked Drain Pump or Filter: A clogged drain pump or filter can prevent water from draining properly, causing the washing machine to skip the spin cycle. This is often accompanied by drainage problems and water remaining in the tub after the wash cycle.

Faulty Motor Coupling: The motor coupling connects the motor to the transmission or drum. If the coupling is worn or broken, the motor may not be able to drive the drum during the spin cycle.

Broken or Worn Drive Motor: A broken or worn drive motor may fail to generate enough power to spin the drum during the spin cycle.

Faulty Control Board: Issues with the control board, such as a malfunctioning motor control module, can prevent the washing machine from entering the spin cycle.

Drum Obstruction: Objects such as coins, small garments, or debris may become lodged between the drum and the tub, preventing the drum from spinning.

To address issues with a washing machine failing to spin:

Balance the Load: Ensure that the load of laundry is evenly distributed around the drum to prevent imbalance during the spin cycle. Avoid overloading the machine with too many clothes.

Check the Lid Switch or Door Lock: Test the lid switch or door lock to ensure it is functioning properly. Replace the switch or lock if it is faulty.

Inspect the Drive Belt: Inspect the drive belt for signs of wear, stretching, or breakage. Replace the belt if it is damaged.

Clean the Drain Pump and Filter: Remove any debris or obstructions from the drain pump and filter to ensure proper drainage. Clean or replace the filter as needed.

Check the Motor Coupling: Inspect the motor coupling for signs of wear or damage. Replace the coupling if it is worn or broken.

Test the Drive Motor: Test the drive motor to ensure it is functioning properly. If the motor is faulty, it may need to be repaired or replaced.

Inspect the Control Board: Check the control board for any error codes or malfunctions. Replace the control board if necessary.

Remove Drum Obstructions: Inspect the drum for any objects or debris that may be obstructing movement. Remove any obstructions to allow the drum to spin freely.

FAILURE TO AGITATE

When a washing machine fails to agitate, it can result in clothes that are not properly cleaned. Several common issues may be causing this problem:

Faulty Agitator Dogs: Agitator dogs are small plastic components that engage with the agitator to create the back-and-forth motion necessary for agitation. If these dogs wear out or break, the agitator may not function properly.

Worn Agitator Cam Kit or Agitator Repair Kit: The agitator cam kit or agitator repair kit contains components that facilitate the movement of the agitator. If these parts wear out or become damaged, the agitator may fail to agitate.

Faulty Drive Belt: The drive belt transfers power from the motor to the transmission or agitator. If the belt is worn, stretched, or broken, the agitator may not receive sufficient power to agitate properly.

Malfunctioning Agitator Motor Coupling: The motor coupling connects the motor to the transmission or agitator. If the coupling is worn, broken, or damaged, the agitator may not receive adequate power to agitate.

Clogged or Faulty Water Inlet Valve: The water inlet valve controls the flow of water into the washing machine. If the valve is clogged or malfunctioning, it may not allow enough water into the machine, preventing proper agitation.

Faulty Timer or Control Board: Issues with the timer or control board can prevent the washing machine from advancing through the wash cycle properly, including the agitation phase.

Loose or Damaged Agitator: If the agitator is loose or damaged, it may not engage with the agitator shaft properly, resulting in ineffective agitation.

Overloaded Washing Machine: Overloading the washing machine with too much laundry can strain the agitator and prevent it from agitating effectively.

To address issues with a washing machine failing to agitate:

Inspect and Replace Agitator Dogs: Check the agitator dogs for wear or damage and replace them if necessary.

Replace Agitator Cam Kit or Agitator Repair Kit: Replace the agitator cam kit or agitator repair kit if the components are worn or damaged.

Inspect and Replace Drive Belt: Check the drive belt for signs of wear, stretching, or damage and replace it if necessary.

Replace Agitator Motor Coupling: Inspect the motor coupling for wear, damage, or breakage and replace it if necessary.

Clean or Replace Water Inlet Valve: Clean the water inlet valve to remove any debris or replace it if it is malfunctioning.

Test and Replace Timer or Control Board: Test the timer or control board for proper function and replace it if necessary.

Inspect and Tighten Agitator: Check the agitator for looseness or damage and tighten or replace it if necessary.

Avoid Overloading: Follow the manufacturer's guidelines for loading the washing machine to prevent overloading and ensure effective agitation.

FOUL ODOR

A foul odor emanating from a washing machine can be unpleasant and often indicates the presence of mold, mildew, or bacteria. Common issues contributing to foul odors in washing machines include:

Residue Buildup: Detergent, fabric softener, and dirt can accumulate over time in various parts of the washing machine, such as the drum, detergent dispenser, and rubber door seal, providing a breeding ground for bacteria and mold.

Standing Water: If water remains trapped in the drum, drain pump, or filter between wash cycles, it can stagnate and emit a foul odor. This may occur due to drainage problems or blockages in the drain system.

Poor Ventilation: Inadequate airflow around the washing machine can promote the growth of mold and mildew. This is particularly common in laundry rooms with limited ventilation or in washing machines installed in enclosed spaces.

Low Washing Temperatures: Washing laundry at low temperatures, such as cold or lukewarm water, may not effectively kill bacteria or mold spores, allowing them to proliferate and cause foul odors.

Using Too Much Detergent: Excessive use of detergent or fabric softener can lead to residue buildup in the washing machine, contributing to foul odors. Using the correct amount of detergent according to the manufacturer's instructions can help prevent this issue.

Infrequent Cleaning: Neglecting to clean the washing machine regularly allows residue, mold, and bacteria to accumulate, leading to foul odors over time.

Using the Wrong Detergent: Using the wrong type of detergent or using detergent formulated for high-efficiency (HE) machines in a non-HE washing machine can result in excess suds and residue buildup, contributing to foul odors.

To address foul odors in a washing machine:

Clean the Drum and Detergent Dispenser: Wipe down the drum, detergent dispenser, and rubber door seal with a solution of water and white vinegar or a mild detergent to remove residue and kill bacteria. Scrub any visible mold or mildew with a soft brush.

Run a Maintenance Wash: Run an empty hot water cycle with a washing machine cleaner or a mixture of white vinegar and baking soda to disinfect the drum and eliminate odors. Follow the manufacturer's instructions for the cleaner or mixture.

Clean the Drain Pump and Filter: Regularly clean the drain pump and filter to remove any debris or standing water that may contribute to foul odors. Refer to the washing machine's manual for instructions on how to access and clean these components.

Ensure Proper Ventilation: Improve airflow around the washing machine by opening windows, installing exhaust fans, or using a dehumidifier in the laundry room to prevent mold and mildew growth.

Wash Laundry at Higher Temperatures: Wash laundry at the hottest temperature recommended for the fabric to kill bacteria and mold spores effectively. Consider running occasional hot water cycles with no laundry to disinfect the machine.

Use the Correct Amount of Detergent: Follow the manufacturer's instructions for the correct amount of detergent to use based on load size and water hardness. Avoid overloading the machine with detergent, as this can lead to residue buildup.

Regular Maintenance: Clean the washing machine regularly, including the drum, detergent dispenser, and rubber door seal, to prevent residue buildup and foul odors. Consider scheduling routine maintenance washes to keep the machine clean and odor free.

ELECTRICAL CONTROL BOARD

The electrical control board in a washing machine is a critical component responsible for controlling various functions, cycles, and operations of the machine. Common issues with the control board include:

Power Surges: Power surges or voltage spikes can damage the control board, leading to malfunctions or complete failure. This can occur during electrical storms or when the washing machine is plugged into an unstable power source.

Water Damage: Water leaks or spills near the control board can cause short circuits or corrosion, damaging the electronic components and affecting the functionality of the control board.

Component Failure: Individual components on the control board, such as relays, capacitors, or resistors, may fail over time due to wear, heat, or manufacturing defects, resulting in erratic behavior or failure of the control board.

Overheating: Excessive heat generated by the washing machine's motor or other components can cause the control board to overheat, leading to damage or failure. Poor ventilation or blocked airflow around the control board can exacerbate this issue.

Corrosion: Corrosion of the electrical contacts or solder joints on the control board can occur over time, especially in humid or damp environments, leading to poor electrical connections and malfunctioning of the board.

Software Glitches: Software glitches or programming errors in the control board's firmware or software can cause erratic behavior, error codes, or unresponsive controls.

Physical Damage: Physical damage to the control board, such as impact or excessive vibration, can cause cracks, breaks, or other damage to the circuitry, resulting in malfunction or failure.

To address issues with the washing machine's electrical control board:

Power Cycle the Machine: Try resetting the washing machine by unplugging it from the power source for a few minutes and then plugging it back in. This can sometimes clear minor software glitches or temporary faults in the control board.

Check for Loose Connections: Inspect the wiring harnesses and connectors on the control board for any loose or disconnected wires. Reconnect any loose wires and ensure all connections are secure.

Inspect for Water Damage: Check for signs of water damage or corrosion on the control board. If water damage is present, the control board may need to be replaced.

Test Components: Use a multimeter to test individual components on the control board, such as relays, capacitors, and resistors, for continuity and proper function. Replace any faulty components as needed.

Inspect for Physical Damage: Visually inspect the control board for any signs of physical damage, such as cracks, burns, or broken components. If physical damage is present, the control board may need to be replaced.

Software Update or Reset: If the washing machine allows for it, try updating or resetting the software or firmware of the control board. Follow the manufacturer's instructions for updating or resetting the software.

Replace the Control Board: If troubleshooting and inspection reveal significant damage or malfunction in the control board, it may need to be replaced. Consult the washing machine's manual or contact the manufacturer for guidance on obtaining a replacement control board and instructions for installation.

WATER TEMPERATURE PROBLEMS

Water temperature problems in a washing machine can lead to ineffective cleaning and rinsing of laundry. Common issues causing water temperature problems include:

Faulty Water Inlet Valve: The water inlet valve controls the flow of hot and cold water into the washing machine. If the valve is defective or clogged, it may not properly regulate the water temperature, leading to issues with hot, cold, or mixed water settings.

Clogged Water Inlet Screens: Sediment or debris can accumulate in the water inlet screens or filters, restricting water flow and affecting temperature control. This can result in insufficient hot or cold water entering the machine.

Incorrect Water Supply Connections: Incorrect plumbing connections or reversed hot and cold water supply lines can lead to incorrect water temperature settings. Ensure that the hot water supply is connected to the hot water inlet and the cold water supply is connected to the cold water inlet on the washing machine.

Sediment Buildup in Water Heater: Sediment buildup in the water heater can affect its performance and efficiency, resulting in inconsistent water temperature output. Flushing the water heater periodically can help remove sediment and improve water temperature control.

Faulty Temperature Sensor: The temperature sensor or thermistor monitors the water temperature inside the washing machine. If the sensor is faulty or malfunctioning, it may not accurately detect or regulate the water temperature, leading to issues with temperature control.

Water Heater Issues: Problems with the home's water heater, such as a malfunctioning thermostat or heating element, can result in inadequate hot water supply to the washing machine, affecting temperature control.

Low Water Pressure: Insufficient water pressure from the hot or cold water supply lines can affect the washing machine's ability to maintain consistent water temperature settings. Check for adequate water pressure from both supply lines.

Plumbing Issues: Issues with the home's plumbing system, such as leaks, blockages, or pipe corrosion, can affect water flow and temperature control. Inspect the plumbing connections and address any issues as needed.

To address water temperature problems in a washing machine:

Check Water Inlet Valve: Inspect the water inlet valve for signs of damage or debris buildup. Clean or replace the valve if necessary.

Clean Water Inlet Screens: Remove and clean the water inlet screens or filters to ensure unrestricted water flow. Replace any damaged or clogged screens.

Verify Water Supply Connections: Ensure that the hot and cold water supply lines are correctly connected to the corresponding inlets on the washing machine.

Flush Water Heater: Periodically flush the water heater to remove sediment buildup and improve water temperature control.

Test Temperature Sensor: Test the temperature sensor or thermistor for proper function using a multimeter. Replace the sensor if it is faulty.

Check Water Heater: Inspect the water heater for issues such as thermostat malfunctions or heating element failures. Repair or replace the water heater components as needed.

Verify Water Pressure: Check the water pressure from both hot and cold water supply lines. Ensure that the pressure meets the manufacturer's recommendations.

Inspect Plumbing: Check for any plumbing issues within the home, such as leaks, blockages, or corrosion. Repair or replace plumbing components as needed.

DETERGENT DISPENSER ISSUES

Detergent dispenser issues in a washing machine can result in improper dispensing of detergent or fabric softener, leading to ineffective cleaning or staining of laundry. Common problems with detergent dispensers include:

Clogged Dispenser Drawer: Accumulated detergent residue or fabric softener can clog the dispenser drawer, preventing proper dispensing during the wash cycle.

Leaking Dispenser: A leaking dispenser can result from damaged seals, cracks, or misalignment, causing detergent or fabric softener to leak onto the clothes or into the washing machine.

Dispenser Drawer Not Closing Properly: If the dispenser drawer does not close properly or is misaligned, it may not dispense detergent or fabric softener correctly during the wash cycle.

Faulty Dispenser Actuator or Solenoid: The dispenser actuator or solenoid controls the release of detergent or fabric softener into the washing machine. If these components are faulty or malfunctioning, the dispenser may not operate correctly.

Incorrect Dispenser Selection: Using the wrong dispenser compartment for the type of detergent or fabric softener being used can lead to improper dispensing or mixing of products.

Buildup of Mold or Mildew: Moisture and detergent residue in the dispenser drawer can create an environment conducive to mold or mildew growth, causing foul odors and potential dispenser clogging.

Damaged or Worn Dispenser Components: Components within the detergent dispenser, such as the syphon or jets, may become damaged or worn over time, affecting proper dispensing.

> ### To address detergent dispenser issues in a washing machine:

Clean the Dispenser Drawer: Remove the dispenser drawer and clean it thoroughly with warm, soapy water to remove detergent residue or fabric softener buildup. Use a brush to scrub any stubborn residue and rinse the drawer thoroughly before reinstalling it.

Clear Clogs: Check for and remove any clogs or blockages in the dispenser drawer or dispenser housing using a soft brush or cloth. Pay attention to small holes or channels where detergent or fabric softener flows.

Inspect and Replace Seals: Check the dispenser drawer and housing for damaged or worn seals that may be causing leaks. Replace any damaged seals to prevent leaks.

Check Dispenser Mechanism: Inspect the dispenser actuator or solenoid for proper function. Ensure that it moves freely and operates correctly during the wash cycle. Replace any faulty components as needed.

Ensure Proper Dispenser Selection: Use the correct dispenser compartment for the type of detergent or fabric softener being used. Follow the manufacturer's recommendations for loading the dispenser drawer.

Prevent Mold and Mildew: Keep the dispenser drawer and housing clean and dry to prevent mold or mildew growth. Leave the dispenser drawer open between wash cycles to allow air circulation and reduce moisture buildup.

Inspect for Damage: Check for any damage or wear to dispenser components, such as the syphon or jets. Replace damaged or worn parts to ensure proper dispensing.

Dryers

FAILURE TO HEAT

When a dryer fails to heat properly, it can result in damp or wet clothes after a drying cycle. Several common issues may be causing this problem:

Faulty Heating Element (Electric Dryers): The heating element is responsible for generating heat in electric dryers. If the element is faulty, damaged, or broken, the dryer won't produce heat.

Thermal Fuse Tripped: The thermal fuse acts as a safety device to prevent the dryer from overheating. If the thermal fuse is tripped due to overheating, it will interrupt the circuit and prevent the heating element from functioning.

Faulty Igniter (Gas Dryers): Gas dryers use an igniter to ignite the gas and produce heat. If the igniter is faulty or not working properly, the gas burner won't ignite, resulting in a lack of heat.

Malfunctioning Gas Valve Solenoid (Gas Dryers): Gas valve solenoids control the flow of gas to the burner. If one or more solenoids fail, the gas burner won't function, and the dryer won't produce heat.

Defective Thermostat: The thermostat regulates the temperature inside the dryer. A defective thermostat may not signal the heating element or gas burner to turn on, leading to a lack of heat.

Blocked Vent or Lint Filter: Restricted airflow due to a clogged lint filter or a blocked vent can cause overheating, leading to the thermal fuse tripping. It can also reduce the efficiency of the dryer and result in inadequate heating.

Faulty High-Limit Thermostat: The high-limit thermostat is another safety device that shuts off the heating element if the dryer overheats. If it malfunctions, it may interrupt the heating process even when the dryer is not overheating.

Broken or Worn Belt (Tumble-Style Dryers): In tumble-style dryers, a broken or worn belt can prevent the drum from turning, leading to a lack of heat.

Issues with Electrical Components: Problems with wiring, switches, or electrical connections in the dryer can disrupt the flow of electricity to the heating element.

To address a dryer's failure to heat:

Check the Heating Element (Electric Dryers): Test the heating element for continuity using a multimeter. Replace the heating element if it is faulty or damaged.

Inspect the Thermal Fuse: Test the thermal fuse for continuity. If it's tripped, investigate the cause of overheating and replace the thermal fuse.

Examine the Igniter and Gas Valve Solenoids (Gas Dryers): Check the igniter for visible damage. If it's faulty, replace it. Additionally, inspect the gas valve solenoids and replace any defective ones.

Test the Thermostat and High-Limit Thermostat: Use a multimeter to check the thermostats for continuity. Replace any thermostats that are faulty or not functioning correctly.

Clear Blocked Vents and Lint Filters: Ensure that the lint filter is clean, and check for blockages in the vent system. Clear any lint or debris to improve airflow.

Inspect the Drum Belt (Tumble-Style Dryers): Check for a broken or worn drum belt in tumble-style dryers. Replace the belt if necessary.

Examine Electrical Components: Inspect wiring, switches, and electrical connections for signs of damage or wear. Replace or repair any faulty components.

POOR DRYING PERFORMANCE

When a dryer has poor drying performance, clothes may come out damp or take longer than usual to dry completely. Several common issues could be causing this problem.

Clogged Lint Filter: A clogged or dirty lint filter restricts airflow, reducing the dryer's efficiency and prolonging drying times. Clean the lint filter before each use to ensure proper airflow.

Blocked Ventilation System: A blocked or restricted dryer vent prevents hot, moist air from escaping the dryer, leading to poor drying performance. Inspect the vent hose and exterior vent opening for blockages, and clean or clear any obstructions.

Overloading: Overloading the dryer with too many clothes can prevent proper airflow and heat distribution, resulting in longer drying times. Avoid overloading the dryer and only dry small to medium-sized loads at a time.

Improper Ventilation Installation: Improperly installed dryer vents with excessive bends or long lengths can impede airflow, causing poor drying performance. Ensure the vent hose is installed correctly and is as short and straight as possible.

Faulty Heating Element: In electric dryers, a malfunctioning heating element can lead to inadequate heat production, resulting in poor drying performance. Test the heating element for continuity using a multimeter and replace it if defective.

Gas Supply Issues (Gas Dryers): Gas dryers require a steady supply of natural gas or propane to produce heat. If there are issues with the gas supply, such as low pressure or a closed valve, the dryer may not dry clothes effectively.

Faulty Thermostat: A faulty thermostat may not accurately regulate the dryer's temperature, leading to inconsistent drying performance. Test the thermostat for continuity and replace it if necessary.

Worn Drum Seal or Gasket: A worn or damaged drum seal or gasket can allow hot air to escape from the dryer drum, reducing drying efficiency. Inspect the seal or gasket for signs of wear and replace it if necessary.

Dirty Dryer Interior: Accumulated lint, debris, or moisture inside the dryer can obstruct airflow and affect drying performance. Clean the interior of the dryer, including the drum, baffles, and exhaust duct, regularly to maintain optimal performance.

Old or Faulty Dryer Vent Hose: Over time, dryer vent hoses can become worn or damaged, leading to leaks or airflow restrictions. Replace the vent hose if it is old, damaged, or leaking.

Faulty Timer or Control Panel: Issues with the dryer's timer or control panel can result in incorrect cycle settings or failure to complete drying cycles. Test the timer and control panel components for proper function and replace them if necessary.

Incorrect Dryer Settings: Using incorrect drying settings or cycle options for the type of load being dried can result in poor drying performance. Select the appropriate drying cycle and settings for the load being dried.

To address poor drying performance in a dryer:

Clean the Lint Filter: Remove any lint or debris from the lint filter before each use to ensure proper airflow.

Check and Clean the Ventilation System: Inspect the dryer vent hose and exterior vent opening for blockages, and clean or clear any obstructions.

Avoid Overloading: Only dry small- to medium-sized loads at a time to prevent overloading the dryer.

Inspect and Test Heating Element: Test the heating element for continuity using a multimeter and replace it if defective.

Ensure Proper Gas Supply (Gas Dryers): Verify that the gas supply to the dryer is adequate and the gas valve is open.

Test Thermostat and Replace if Faulty: Test the thermostat for continuity and replace it if necessary to ensure accurate temperature regulation.

Clean the Dryer Interior: Remove lint, debris, and moisture from the interior of the dryer regularly to maintain optimal airflow.

Inspect and Replace Worn Seals or Gaskets: Inspect the drum seal or gasket for wear or damage and replace it if necessary to prevent heat loss.

Replace Old or Faulty Vent Hose: Replace the dryer vent hose if it is old, damaged, or leaking to ensure proper airflow.

Check Timer and Control Panel Settings: Verify that the dryer's timer and control panel settings are correct for the type of load being dried.

NOISY OPERATION

Noisy operation in a dryer can be disruptive and may indicate several potential issues with the appliance. Here are some common causes of noisy dryer operation:

Worn Drum Support Rollers: Drum support rollers support the dryer drum as it rotates. Over time, these rollers can wear out or develop flat spots, causing a thumping or rumbling noise during operation.

Faulty Drum Bearing: The drum bearing supports the rear of the dryer drum. If the bearing wears out or becomes damaged, it can result in a squealing or grinding noise as the drum rotates.

Worn Drum Glides or Pads: Drum glides or pads provide a smooth surface for the dryer drum to rotate on. If these components wear out, the drum may rub against the dryer cabinet, causing a scraping or squeaking noise.

Loose or Worn Belt: The drive belt that rotates the dryer drum can become loose or worn out over time. A loose or damaged belt can cause a thumping or squealing noise as it moves around the drum pulley.

Misaligned Drum: If the dryer drum becomes misaligned or off-center, it may rub against other components or the dryer cabinet, resulting in noise during operation.

Foreign Objects in the Drum: Coins, buttons, or other small objects can become lodged between the dryer drum and the drum baffles, causing a rattling or banging noise during operation.

Worn Drum Seal: The drum seal or gasket creates a seal between the dryer drum and the dryer cabinet. If the seal becomes worn or damaged, it can result in excess friction and noise as the drum rotates.

Damaged Blower Wheel: The blower wheel circulates air inside the dryer. If the blower wheel becomes damaged or clogged with debris, it can produce a loud rattling or vibrating noise.

Faulty Motor Bearings: The dryer motor contains bearings that can wear out over time, resulting in a loud humming or grinding noise during operation.

Loose or Worn Idler Pulley: The idler pulley provides tension to the dryer belt. If the pulley becomes loose or worn, it can cause a squealing or rattling noise as the dryer drum rotates.

To address noisy operation in a dryer:

Inspect Drum Support Rollers, Bearings, and Glides: Check these components for signs of wear or damage and replace them if necessary.

Tighten or Replace the Drive Belt: Ensure the drive belt is properly tensioned and replace it if it is loose or worn.

Check for Foreign Objects: Remove any foreign objects lodged between the dryer drum and the drum baffles.

Inspect and Replace the Drum Seal: If the drum seal is worn or damaged, replace it to reduce friction and noise.

Clean or Replace the Blower Wheel: Remove any debris or obstructions from the blower wheel and replace it if it is damaged.

Inspect and Lubricate Motor Bearings: If the motor bearings are worn or damaged, replace the motor or lubricate the bearings if possible.

Tighten or Replace the Idler Pulley: Ensure the idler pulley is properly tensioned and replace it if it is loose or worn.

Ensure the Dryer Is Level: Level the dryer to prevent misalignment and reduce noise during operation.

DRUM NOT TURNING

When a dryer drum fails to turn during operation, it can prevent clothes from drying properly. Several common issues could be causing this problem.

Broken Drive Belt: The drive belt is responsible for rotating the dryer drum. If the belt is broken or slipped off the pulley, the drum won't turn. Inspect the belt for signs of damage or wear and replace it if necessary.

Worn Drum Rollers or Bearings: Drum support rollers or bearings support the dryer drum as it rotates. If these components are worn out or damaged, they can prevent the drum from turning smoothly. Check the rollers or bearings for signs of wear or damage and replace them if needed.

Faulty Drum Motor: The drum motor powers the rotation of the dryer drum. If the motor is defective or burned out, it won't be able to turn the drum. Test the motor for continuity using a multimeter and replace it if necessary.

Broken Drum Belt Switch: Many dryers are equipped with a safety switch that prevents the dryer from operating if the drive belt is broken. If this switch is faulty or defective, it may prevent the drum from turning even if the belt is intact.

Blocked Blower Wheel: The blower wheel circulates air inside the dryer. If the blower wheel is blocked by lint or debris, it may prevent the drum from turning. Inspect the blower wheel and clean any obstructions.

Faulty Start Switch or Door Switch: The start switch and the door switch activate the dryer when the door is closed and the start button is pressed. If these switches are faulty or defective, they may prevent the drum from turning. Test the switches for continuity and replace them if necessary.

Overloaded Drum: Overloading the dryer drum with too many clothes can strain the drive motor and prevent the drum from turning. Reduce the load size and try again.

Foreign Objects in the Drum: Objects such as coins, buttons, or small articles of clothing can become lodged between the drum and the dryer cabinet, preventing the drum from turning. Check for and remove any foreign objects.

To address a dryer drum that isn't turning:

Inspect and Replace the Drive Belt: Check the drive belt for signs of damage or wear and replace it if necessary.

Check Drum Rollers or Bearings: Inspect the drum support rollers or bearings for signs of wear or damage and replace them if needed.

Test the Drum Motor: Use a multimeter to test the drum motor for continuity and replace it if defective.

Check the Drum Belt Switch: Ensure that the drum belt switch is functioning properly and replace it if defective.

Inspect the Blower Wheel: Check the blower wheel for obstructions and clean any lint or debris.

Test the Start Switch and Door Switch: Test the start switch and the door switch for continuity and replace them if necessary.

Reduce the Load Size: Avoid overloading the dryer drum with too many clothes.

Remove Foreign Objects: Check for and remove any foreign objects lodged between the drum and the dryer cabinet.

OVERHEATING AND BURNING SMELL

When a dryer is overheating and emitting a burning smell, it's crucial to address the issue promptly to prevent potential fire hazards. Several common issues could be causing this problem.

Lint Buildup: Accumulated lint inside the dryer or the exhaust vent can obstruct airflow and cause the dryer to overheat. This can result in a burning smell and pose a fire hazard. Clean the lint trap after every use and regularly inspect and clean the exhaust vent to prevent lint buildup.

Blocked Ventilation System: A blocked or restricted dryer vent can prevent hot air from escaping, leading to overheating and a burning smell. Inspect the vent hose and exterior vent opening for blockages, and clean or clear any obstructions.

Faulty Heating Element or Gas Burner Assembly: A malfunctioning heating element (in electric dryers) or gas burner assembly (in gas dryers) can cause the dryer to overheat. This can result in overheating and a burning smell. Inspect the heating element or gas burner assembly for signs of damage or malfunction and replace them if necessary.

Thermostat Malfunction: A faulty thermostat may fail to regulate the dryer's temperature properly, causing it to overheat. Test the thermostat for continuity and replace it if defective.

Blocked Dryer Cabinet: Blockages or obstructions inside the dryer cabinet can restrict airflow and cause the dryer to overheat. Inspect the dryer cabinet for any obstructions and clear them if necessary.

Worn Drum Seal or Gasket: A worn or damaged drum seal or gasket can allow hot air to escape from the dryer drum, leading to overheating and a burning smell. Inspect the seal or gasket for signs of wear and replace it if necessary.

Faulty High-Limit Thermostat or Thermal Fuse: The high-limit thermostat and thermal fuse are safety devices that shut off the heating element if the dryer overheats. If these components malfunction, they may fail to shut off the heating element, causing overheating and a burning smell. Test the high-limit thermostat and thermal fuse for continuity and replace them if defective.

Foreign Objects in the Dryer: Objects such as clothing labels, buttons, or other debris can become lodged inside the dryer and cause friction, leading to overheating and a burning smell. Inspect the dryer drum and remove any foreign objects.

To address a dryer overheating and emitting a burning smell:

Clean the Lint Trap and Exhaust Vent: Remove any lint or debris from the lint trap after every use and regularly clean the exhaust vent to prevent lint buildup.

Inspect and Clean the Ventilation System: Check the vent hose and exterior vent opening for blockages and clear any obstructions to ensure proper airflow.

Check and Replace the Heating Element or Gas Burner Assembly: Inspect the heating element (in electric dryers) or gas burner assembly (in gas dryers) for signs of damage or malfunction and replace them if necessary.

Test and Replace Faulty Thermostats: Test the thermostat and high-limit thermostat for continuity and replace them if defective.

Inspect and Replace Worn Drum Seals or Gaskets: Inspect the drum seal or gasket for signs of wear and replace them if necessary.

Clear Obstructions from the Dryer Cabinet: Check the dryer cabinet for any obstructions or blockages and clear them if necessary.

Remove Foreign Objects from the Dryer Drum: Inspect the dryer drum for foreign objects and remove them to prevent friction and overheating.

TIMER OR CONTROL PANEL ISSUE

Issues with the dryer timer or control panel can lead to various problems with the operation and functionality of the appliance. Some common issues with the dryer timer or control panel include:

Failure to Start: If the dryer fails to start when the start button is pressed, the timer or control panel may be faulty. This could be due to a malfunctioning start switch or a problem with the timer or control board.

Inaccurate Cycle Selection: The timer or control panel may fail to accurately select or advance through drying cycles. This could result in the dryer not completing the selected cycle or running longer than necessary.

Cycle Not Advancing: In some cases, the dryer may start but fail to advance through the selected drying cycle. This could indicate a problem with the timer or control board, preventing the dryer from transitioning to the next phase of the cycle.

Stuck or Frozen Controls: Buttons or knobs on the control panel may become stuck or unresponsive, making it difficult or impossible to select drying options or adjust settings.

Display Errors or Malfunctions: If the dryer has a digital display, errors or malfunctions may occur, such as flashing lights, error codes, or blank displays. These issues could indicate a problem with the control board or display panel.

Intermittent Operation: The dryer may exhibit intermittent operation, where it works sporadically or unpredictably. This could be due to loose connections, wiring issues, or a failing timer or control board.

Noisy Controls: The timer or control panel may produce unusual noises when buttons are pressed or knobs are turned, indicating potential mechanical or electrical problems.

Burnt Smell or Visible Damage: A burnt smell or visible signs of damage, such as melted components or scorched areas, may indicate an electrical problem with the timer or control panel.

To address issues with the dryer timer or control panel:

Reset the Dryer: Try resetting the dryer by unplugging it from the power source for a few minutes and then plugging it back in. This can sometimes clear temporary glitches or errors.

Check for Loose Connections: Inspect the wiring connections between the timer or control panel and other components of the dryer. Tighten any loose connections and ensure proper electrical connections.

Clean and Inspect Controls: Clean the control panel and buttons to remove any dirt, debris, or residue that may be affecting their operation. Inspect for signs of damage or wear.

Test Components with a Multimeter: Use a multimeter to test the continuity of the timer, start switch, and other control panel components. Replace any components that fail the continuity test.

Replace the Timer or Control Board: If troubleshooting steps fail to resolve the issue, consider replacing the timer or control board. Be sure to purchase the correct replacement part compatible with your dryer model.

LINT BUILDUP

Lint buildup in dryers is a common issue that can lead to various problems, including reduced efficiency, increased risk of fire, and potential damage to the dryer. Here are some common issues associated with lint buildup in dryers:

Reduced Drying Efficiency: As lint accumulates in the lint trap, exhaust vent, and dryer ducts, it restricts airflow. Restricted airflow prevents hot, moist air from escaping the dryer efficiently, resulting in longer drying times and reduced overall drying efficiency.

Increased Risk of Fire: Lint is highly flammable, and when it accumulates in large quantities within the dryer and exhaust system, it can pose a significant fire hazard. The heat generated during the drying process can ignite the lint, leading to a dryer fire.

Overheating and Thermal Fuse Failure: Restricted airflow caused by lint buildup can cause the dryer to overheat. In electric dryers, this can lead to thermal fuse failure, which is a safety mechanism designed to shut off the dryer if it overheats. When the thermal fuse fails, the dryer may no longer operate until the fuse is replaced.

Damage to Heating Elements and Components: Overheating caused by lint buildup can also damage heating elements and other internal components of the dryer. Excessive heat can cause components to degrade prematurely, leading to costly repairs or replacements.

Musty Odors: Lint accumulation can trap moisture inside the dryer, creating an ideal environment for mold and mildew growth. This can result in musty odors emanating from the dryer and transferred onto freshly dried laundry.

Blocked Ventilation System: Lint buildup in the exhaust vent and ductwork can block airflow, causing hot, moist air to back up into the dryer. This can lead to condensation, moisture damage, and potential mold growth within the dryer and ventilation system.

Increased Energy Consumption: A dryer with restricted airflow due to lint buildup requires more energy to operate efficiently. This can lead to higher utility bills and increased energy consumption over time.

To address issues with lint buildup in dryers:

Regularly Clean the Lint Trap: Clean the lint trap before or after every load of laundry to prevent lint buildup. Use a lint brush or vacuum attachment to remove any accumulated lint.

Clean the Exhaust Vent and Ductwork: Periodically clean the exhaust vent and ductwork to remove lint buildup. Disconnect the vent from the dryer and vacuum out any lint or debris. If the vent is particularly dirty or clogged, consider hiring a professional vent cleaning service.

Inspect and Clean the Dryer Interior: Remove the dryer's front or rear panel to access the interior components. Use a vacuum or lint brush to remove any lint accumulation from the drum, drum seals, and other internal surfaces.

Check for Lint Behind the Dryer: Occasionally move the dryer away from the wall and check for lint accumulation behind the appliance. Vacuum or sweep behind the dryer to remove any lint buildup.

Install a Lint Trap Brush or Lint Alarm: Consider installing a lint trap brush or lint alarm to help prevent lint buildup and alert you to potential obstructions in the dryer vent system.

DOOR SWITCH MALFUNCTION

A malfunctioning door switch in a dryer can lead to several issues with its operation. Here are some common problems associated with a faulty door switch:

Dryer Won't Start: One of the primary functions of the door switch is to signal to the dryer that the door is securely closed. If the door switch fails, the dryer may not start at all, as it won't receive the signal that the door is closed.

Dryer Stops Mid-Cycle: In some cases, the dryer may start but then stop abruptly during a drying cycle if the door switch is malfunctioning. This occurs because the dryer may detect that the door is open when it's actually closed, causing it to halt operation.

Dryer Runs with Door Open: Conversely, a faulty door switch may fail to recognize when the door is open, allowing the dryer to continue running even when the door is ajar. This poses a safety risk and can lead to accidents if someone inadvertently reaches into the dryer while it's operating.

No Drum Light: Many dryers are equipped with an interior drum light that turns on when the door is opened. If the door switch is malfunctioning, the drum light may not activate, making it difficult to see inside the dryer when loading or unloading laundry.

No Signal to Heating Element or Motor: The door switch is often part of the circuit that supplies power to the heating element and motor. If the switch fails, it may interrupt this circuit, preventing the dryer from heating or tumbling.

Repeated Opening and Closing of the Door: In some cases, a malfunctioning door switch may cause the dryer to repeatedly cycle between thinking the door is open and closed. This can result in the dryer repeatedly starting and stopping or the door latch mechanism clicking continuously.

Visible Damage or Wear on the Switch: Physical damage or wear to the door switch, such as broken or loose components, corrosion, or burnt contacts, may indicate a malfunctioning switch that needs to be replaced.

To address issues with a malfunctioning door switch in a dryer:

Check the Door Alignment: Ensure that the dryer door is properly aligned and closes securely. Sometimes, misalignment or an obstruction can prevent the door from making proper contact with the switch.

Inspect the Door Switch Assembly: Carefully inspect the door switch and its surrounding components for any signs of damage, wear, or corrosion. If the switch appears damaged or worn, it may need to be replaced.

Test the Continuity of the Switch: Use a multimeter to test the continuity of the door switch. With the dryer unplugged, remove the switch from the dryer and test for continuity while pressing the switch button. If there is no continuity when the switch is pressed, it is likely faulty and needs to be replaced.

Replace the Door Switch: If testing confirms that the door switch is faulty, disconnect the wiring harness and remove the switch from the dryer. Install a new door switch that is compatible with your dryer model, reconnect the wiring harness, and reassemble the dryer.

GAS IGNITION PROBLEMS

Gas dryer ignition issues can prevent the appliance from heating properly, leading to incomplete drying cycles or no heat at all. Here are some common problems associated with gas dryer ignition:

Igniter Failure: The igniter is responsible for generating the heat needed to ignite the gas in the burner assembly. If the igniter fails to glow or does not reach the necessary temperature, the gas valve will not open, and the dryer will not heat.

Gas Valve Solenoid Failure: The gas valve solenoids control the flow of gas to the burner assembly. If one or more solenoids fail, the gas valve may not open properly, preventing gas from reaching the burner assembly and igniting.

Clogged Burner Assembly: The burner assembly may become clogged with lint, dirt, or debris over time, obstructing the flow of gas and preventing ignition. A clogged burner assembly can cause erratic heating or no heat at all.

Faulty Flame Sensor: The flame sensor detects the presence of a flame once the gas is ignited. If the flame sensor is faulty or malfunctioning, it may fail to detect the flame, causing the gas valve to shut off prematurely and the dryer to stop heating.

Dirty or Faulty Ignition Sensor: Some gas dryers are equipped with an ignition sensor that detects the presence of a flame during the ignition process. If the sensor is dirty or faulty, it may fail to detect the flame, leading to ignition failure.

Blocked Ventilation System: A blocked or restricted ventilation system can cause overheating inside the dryer, leading to ignition failure or safety shutdowns. Blocked vents can also prevent proper airflow, causing the dryer to overheat and trip the high-limit thermostat.

Gas Supply Issues: Problems with the gas supply, such as low gas pressure or a closed gas valve, can prevent the burner assembly from receiving an adequate supply of gas, leading to ignition failure.

Faulty Control Board: In some cases, a malfunctioning control board can cause ignition issues by failing to send the proper signals to the igniter, gas valve, or other components of the dryer ignition system.

To address issues with gas dryer ignition:

Inspect and Clean the Igniter: Check the igniter for signs of damage or wear and clean it if necessary. If the igniter is faulty, it will need to be replaced.

Test the Gas Valve Solenoids: Use a multimeter to test the continuity of the gas valve solenoids. If any solenoid fails the continuity test, replace them as a set.

Clean the Burner Assembly: Remove any lint, dirt, or debris from the burner assembly using a brush or compressed air. Ensure that the burner ports are clear and unobstructed.

Check the Flame Sensor and the Ignition Sensor: Test the flame sensor and the ignition sensor for continuity using a multimeter. Clean or replace the sensors if necessary.

Inspect the Ventilation System: Ensure that the dryer vent hose, exhaust vent, and lint trap are clean and free of obstructions. Clean or clear any blockages to ensure proper airflow.

Verify the Gas Supply: Check that the gas supply valve is open and that the dryer is receiving an adequate supply of gas. If gas pressure is low, contact a qualified technician to address the issue.

Test the Control Board: If all other components are functioning properly, but the dryer still fails to ignite, the control board may be faulty and require replacement.

Yardi

Creating purchase orders (POs) in Yardi Voyager typically involves the following steps:

Access Voyager: Log in to your Yardi Voyager account with your credentials.

Navigate to Procurement Module: Locate and navigate to the Procurement module within Voyager. This module is where you'll manage purchasing-related activities, including creating and managing purchase orders.

Create a New Purchase Order: Once you're in the Procurement module, look for an option or button to create a new purchase order. This is usually prominently displayed on the interface.

Enter Vendor Information: Select the vendor you want to create the purchase order for and enter their information into the system. If the vendor is not already in the system, you may need to add them as a new vendor.

Add Items: Specify the items you want to order from the vendor. Enter details such as item descriptions, quantities, prices, and any other relevant information.

Review and Approve: Review the purchase order to ensure all information is accurate and complete. Once you're satisfied, submit the purchase order for approval. Depending on your organization's workflow, this may involve sending it to a supervisor or another authorized individual for review and approval.

Finalize and Transmit: After the purchase order is approved, finalize it within the system. This may involve generating a final version of the purchase order and transmitting it electronically to the vendor.

Track and Manage: Throughout the procurement process, use Voyager to track and manage the status of your purchase orders. You can monitor when orders are sent to vendors, track delivery dates, and reconcile invoices against purchase orders.

Receive Goods and Services: Once the vendor delivers the goods or services specified in the purchase order, update the status of the order within Voyager to indicate that it has been received.

Close and Archive: Once the purchase order process is complete, close and archive the purchase order within Voyager for record-keeping purposes.

ORDERING FROM THE MARKETPLACE

To order from the marketplace in Yardi Voyager, you typically follow these steps:

Access Voyager: Log in to your Yardi Voyager account with your credentials.

Navigate to Procurement Module: Locate and navigate to the Procurement module within Voyager. This module is where you manage purchasing-related activities, including ordering from the marketplace.

Access the Marketplace: Look for an option or tab within the Procurement module that allows you to access the marketplace. This may be labeled as "Marketplace" or "Vendor Catalog" depending on how your system is configured.

Browse or Search for Items: Once you're in the marketplace, you can browse or search for the items you want to order. The marketplace typically contains a catalog of products and services available from approved vendors.

Select Items: Select the items you want to order by adding them to your shopping cart or selecting them directly from the catalog.

Review Cart: Review the items in your shopping cart to ensure they are correct and complete. You may have the option to adjust quantities or remove items if needed.

Proceed to Checkout: When you're ready to place your order, proceed to the checkout process. This may involve confirming quantities, selecting shipping options, and providing any additional information required.

Submit Order: Once you've reviewed and confirmed your order details, submit the order to the marketplace vendor for processing. Depending on your organization's workflow, this may also involve obtaining approvals from supervisors or other authorized individuals.

Track Order: After placing your order, use Voyager to track the status of your order. You can monitor when the order is processed, shipped, and delivered.

USING THE MAKE READY BOARD

The Make Ready Board in Yardi Voyager is a tool used in property management to track and manage the process of preparing units for new tenants. Here's how you can use the Make Ready Board:

Access Voyager: Log in to your Yardi Voyager account with your credentials.

Navigate to Make Ready Board: Locate and navigate to the Make Ready Board within Voyager. This feature is typically found in the Maintenance or Property Management module, depending on your organization's configuration.

View Units: Once you're in the Make Ready Board, you'll see a list of units that are currently undergoing the make-ready process. This may include vacant units that need cleaning, repairs, or other maintenance tasks before they can be leased to new tenants.

Track Progress: Each unit listed on the Make Ready Board will have its status displayed, indicating where it is in the make-ready process. Common statuses may include "In Progress," "Ready for Inspection," "Pending Approval," or "Completed."

Assign Tasks: Assign tasks to maintenance staff or contractors responsible for preparing the units. You can assign specific tasks to individuals or teams and set deadlines for completion.

Update Status: As tasks are completed, update the status of each unit on the Make Ready Board accordingly. This helps keep everyone involved in the process informed about the progress being made.

Schedule Inspections: Once the unit is ready for inspection, schedule inspections with property managers or other relevant personnel. After the inspection is completed, update the status of the unit based on the results.

Coordinate Move-In: Once the unit has passed inspection and any necessary approvals have been obtained, coordinate the move-in process with leasing staff and new tenants. Update the status of the unit to indicate that it is ready for occupancy.

Monitor Performance: Use the Make Ready Board to monitor the performance of your maintenance team and track the time it takes to prepare units for new tenants. Identify any bottlenecks or areas for improvement in the make-ready process.

Generate Reports: Utilize reporting features within Voyager to generate reports on make-ready performance, such as average turnaround time, maintenance costs, and unit occupancy rates.

WORK ORDERS

In Yardi Voyager, managing work orders involves several steps, including viewing, updating, and closing them. Here's a general guide on how to see and close work orders:

Access Voyager: Log in to your Yardi Voyager account with your credentials.

Navigate to the Maintenance Module: Work orders are typically managed within the Maintenance module. Locate and navigate to this module within Voyager.

View Work Orders: Once you're in the Maintenance module, you'll typically see a list of all active work orders. These could be sorted by various criteria such as property, unit, status, or priority. Locate the specific work order you want to view or close.

Review Work Order Details: Click on the desired work order to view its details. This will typically include information such as the work order number, description of the issue or task, assigned technician or vendor, status, priority, and any relevant notes or comments.

Update Work Order Status: If the work order has been completed or resolved, update its status to reflect this. Depending on your organization's workflow and configuration, you may have options to change the status to "Completed," "Closed," or another appropriate designation.

Enter Completion Details: If required, enter any additional details or comments related to the completion of the work order. This could include information about the work performed, materials used, or any follow-up actions needed.

Close Work Order: Once all necessary information has been entered and the work order is ready to be closed, select the option to close or finalize the work order. This action typically confirms that the work has been completed to satisfaction and updates the status accordingly.

Generate Reports: After closing the work order, you may want to generate reports to track maintenance performance, analyze trends, and identify areas for improvement. Utilize reporting features within Voyager to generate relevant reports based on your organization's needs.

Archive or Retain Records: Depending on your organization's policies and procedures, you may need to archive or retain records of closed work orders for future reference or auditing purposes. Follow your organization's guidelines for record-keeping and data retention.

Monitor and Follow Up: Continuously monitor the status of work orders to ensure timely completion and customer satisfaction. Follow up with technicians, vendors, or tenants as needed to address any outstanding issues or concerns.

OneSite

MAKING POS

To create purchase orders (POs) in OneSite, you typically follow these steps:

Access OneSite: Log in to your OneSite account with your credentials.

Navigate to the Purchasing Module: Locate and navigate to the Purchasing module within OneSite. This is where you'll manage purchasing-related activities, including creating and managing purchase orders.

Create a New Purchase Order: Once you're in the Purchasing module, look for an option or button to create a new purchase order. This is usually prominently displayed on the interface.

Enter Vendor Information: Select the vendor you want to create the purchase order for and enter their information into the system. If the vendor is not already in the system, you may need to add them as a new vendor.

Add Items: Specify the items you want to order from the vendor. Enter details such as item descriptions, quantities, prices, and any other relevant information.

Review and Approve: Review the purchase order to ensure all information is accurate and complete. Once you're satisfied, submit the purchase order for approval. Depending on your organization's workflow, this may involve sending it to a supervisor or another authorized individual for review and approval.

Finalize and Transmit: After the purchase order is approved, finalize it within the system. This may involve generating a final version of the purchase order and transmitting it electronically to the vendor.

Track and Manage: Throughout the procurement process, use OneSite to track and manage the status of your purchase orders. You can monitor when orders are sent to vendors, track delivery dates, and reconcile invoices against purchase orders.

Receive Goods and Services: Once the vendor delivers the goods or services specified in the purchase order, update the status of the order within OneSite to indicate that it has been received.

Close and Archive: Once the purchase order process is complete, close and archive the purchase order within OneSite for record-keeping purposes.

ORDERING FROM THE MARKETPLACE

To order from the marketplace in OneSite, you typically follow these steps:

Access OneSite: Log in to your OneSite account with your credentials.

Navigate to the Procurement Module: Locate and navigate to the Procurement module within OneSite. This module is where you manage purchasing-related activities, including ordering from the marketplace.

Access the Marketplace: Look for an option or tab within the Procurement module that allows you to access the marketplace. This may be labeled as "Marketplace" or "Vendor Catalog" depending on how your system is configured.

Browse or Search for Items: Once you're in the marketplace, you can browse or search for the items you want to order. The marketplace typically contains a catalog of products and services available from approved vendors.

Select Items: Select the items you want to order by adding them to your shopping cart or selecting them directly from the catalog.

Review Cart: Review the items in your shopping cart to ensure they are correct and complete. You may have the option to adjust quantities or remove items if needed.

Proceed to Checkout: When you're ready to place your order, proceed to the checkout process. This may involve confirming quantities, selecting shipping options, and providing any additional information required.

Submit Order: Once you've reviewed and confirmed your order details, submit the order to the marketplace vendor for processing. Depending on your organization's workflow, this may also involve obtaining approvals from supervisors or other authorized individuals.

Track Order: After placing your order, use OneSite to track the status of your order. You can monitor when the order is processed, shipped, and delivered.

THE MAKE READY BOARD

In OneSite, the Make Ready Board is a tool used in property management to track and manage the process of preparing units for new tenants. Here's how you can use the Make Ready Board:

Access OneSite: Log in to your OneSite account with your credentials.

Navigate to the Make Ready Board: Locate and navigate to the Make Ready Board within OneSite. This feature is typically found in the Maintenance or Property Management module, depending on your organization's configuration.

View Units: Once you're in the Make Ready Board, you'll typically see a list of units that are currently undergoing the make-ready process. These could be vacant units that need cleaning, repairs, or other maintenance tasks before they can be leased to new tenants.

Track Progress: Each unit listed on the Make Ready Board will have its status displayed, indicating where it is in the make-ready process. Common statuses may include "In Progress," "Ready for Inspection," "Pending Approval," or "Completed."

Assign Tasks: Assign tasks to maintenance staff or contractors responsible for preparing the units. You can assign specific tasks to individuals or teams and set deadlines for completion.

Update Status: As tasks are completed, update the status of each unit on the Make Ready Board accordingly. This helps keep everyone involved in the process informed about the progress being made.

Schedule Inspections: Once the unit is ready for inspection, schedule inspections with property managers or other relevant personnel. After the inspection is completed, update the status of the unit based on the results.

Coordinate Move-In: Once the unit has passed inspection and any necessary approvals have been obtained, coordinate the move-in process with leasing staff and new tenants. Update the status of the unit to indicate that it is ready for occupancy.

Monitor Performance: Use the Make Ready Board to monitor the performance of your maintenance team and track the time it takes to prepare units for new tenants. Identify any bottlenecks or areas for improvement in the make-ready process.

Generate Reports: After using the Make Ready Board, you may want to generate reports to track maintenance performance, analyze trends, and identify areas for improvement. Utilize reporting features within OneSite to generate relevant reports based on your organization's needs.

WORK ORDERS

In OneSite, managing work orders involves several steps, including viewing, updating, and closing them. Here's a general guide on how to see and close work orders:

Access OneSite: Log in to your OneSite account with your credentials.

Navigate to the Work Orders Module: Locate and navigate to the Work Orders module within OneSite. This is where you'll manage work orders related to maintenance and repairs.

View Work Orders: Once you're in the Work Orders module, you'll typically see a list of all active work orders. These could be sorted by various criteria such as property, unit, status, or priority. Locate the specific work order you want to view or close.

Review Work Order Details: Click on the desired work order to view its details. This will typically include information such as the work order number, description of the issue or task, assigned technician or vendor, status, priority, and any relevant notes or comments.

Update Work Order Status: If the work order has been completed or resolved, update its status to reflect this. Depending on your organization's workflow and configuration, you may have options to change the status to "Completed," "Closed," or another appropriate designation.

Enter Completion Details: If required, enter any additional details or comments related to the completion of the work order. This could include information about the work performed, materials used, or any follow-up actions needed.

Close Work Order: Once all necessary information has been entered and the work order is ready to be closed, select the option to close or finalize the work order. This action typically confirms that the work has been completed to satisfaction and updates the status accordingly.

Generate Reports: After closing the work order, you may want to generate reports to track maintenance performance, analyze trends, and identify areas for improvement. Utilize reporting features within OneSite to generate relevant reports based on your organization's needs.

Archive or Retain Records: Depending on your organization's policies and procedures, you may need to archive or retain records of closed work orders for future reference or auditing purposes. Follow your organization's guidelines for recordkeeping and data retention.

Monitor and Follow Up: Continuously monitor the status of work orders to ensure timely completion and customer satisfaction. Follow up with technicians, vendors, or tenants as needed to address any outstanding issues or concerns.

Microwave

ELECTRICAL PROBLEMS

Power Supply Failure: The microwave may fail to turn on due to issues with the power supply. This could be caused by a blown fuse, tripped circuit breaker, or a problem with the outlet itself.

Intermittent Power Loss: The microwave might experience intermittent power loss, where it turns off unexpectedly during operation. This could be due to loose wiring connections, a faulty power cord, or issues with the control board.

Control Panel Failure: Problems with the control panel, such as unresponsive buttons or erratic behavior, can occur due to electrical issues within the panel's circuitry.

Blown Fuse: A blown fuse can result from power surges or electrical faults within the microwave. It's a safety feature designed to protect the appliance from damage, but repeated fuse blows indicate an underlying problem that needs to be addressed.

Short Circuits: Short circuits can occur when wires come into contact with each other or with metal components inside the microwave. This can cause electrical arcing, sparks, and potentially lead to fires if not addressed.

Overheating Components: Electrical components inside the microwave, such as the magnetron, capacitor, or transformer, can overheat due to electrical faults or excessive load. Overheating can cause these components to fail prematurely.

High Voltage Diode Failure: The high voltage diode is responsible for converting the microwave's power supply into the high voltage required by the magnetron. If the diode fails, the microwave may not heat properly or may not work at all.

Faulty Magnetron: The magnetron generates the microwaves used for heating food. Electrical problems within the magnetron can lead to uneven heating, no heating, or sparking inside the microwave.

Grounding Issues: Improper grounding of the microwave can lead to electrical problems, including electrical shocks, equipment damage, and increased risk of electrical fires.

Capacitor Issues: Capacitors store electrical energy and help regulate the flow of electricity within the microwave. Capacitor failure can lead to various issues, including erratic behavior, buzzing sounds, or even dangerous electrical discharge.

> **Repairing electrical issues in microwaves can be complex and potentially hazardous due to the high voltages involved. It's essential to prioritize safety.**

Safety Precautions: Unplug the microwave from the power source before attempting any repairs. Use insulated tools and wear appropriate personal protective equipment (PPE) such as gloves and safety glasses. Be aware of the high-voltage components inside the microwave, such as the capacitor and magnetron, which can store dangerous electrical charges even when unplugged.

Diagnostic Steps: Identify the specific symptoms and behaviors of the microwave to narrow down potential causes. For example, does the microwave not turn on at all, or does it turn on but not heat food? Inspect the power cord, plug, and outlet for any signs of damage or loose connections. Check the microwave's fuse and replace it if blown. However, if the fuse blows repeatedly, it indicates an underlying issue that needs further investigation.

Visual Inspection: Carefully examine the internal components of the microwave for signs of damage, such as burnt wires, charred components, or loose connections. Look for any signs of overheating, such as discoloration or melted insulation around electrical components.

Testing Components: Use a multimeter to test the continuity of electrical components, such as the fuse, diode, capacitor, and magnetron. Check for proper voltage levels at various points in the circuit to ensure that power is reaching the appropriate components.

Repair or Replacement: Replace any damaged or faulty components identified during the testing phase. Follow manufacturer's guidelines and use OEM (original equipment manufacturer) replacement parts whenever possible to ensure compatibility and safety. If you're unsure about the repair process or lack the necessary expertise, it's best to consult a professional technician or contact the manufacturer for assistance.

Reassembly and Testing: Reassemble the microwave carefully, ensuring that all components are properly connected and secured. Plug in the microwave and perform a test run to verify that the issue has been resolved and that the appliance functions correctly.

UNUSUAL NOISES

Unusual noises in a microwave can indicate various issues with different components. Here are some common problems associated with unusual noises:

Turntable Issues: If you hear grinding, scraping, or rattling noises, it might indicate problems with the turntable or its support mechanism. Check if the turntable is properly seated on its support and if there are any obstructions hindering its movement.

Magnetron: A buzzing or humming noise may suggest issues with the magnetron, the component responsible for generating microwave radiation. If the magnetron is failing or malfunctioning, it can produce unusual sounds during operation.

Fan Motor: Most microwaves have a fan that helps cool the internal components and circulate air during operation. If the fan motor is malfunctioning or obstructed, it can produce loud or irregular noises.

Rotating Ring or Support: If your microwave has a rotating ring or support mechanism for the turntable, noises may occur if these components are misaligned, damaged, or worn out.

High Voltage Diode or Capacitor: Faulty high-voltage components like the diode or capacitor can lead to unusual noises, such as clicking or buzzing sounds. These components play crucial roles in converting the power supply to the high voltage needed for the magnetron.

Exhaust System: If your microwave has an exhaust system for venting steam and odors, noises may occur due to obstructions or issues with the exhaust fan or motor.

Door Components: Issues with the door or its components, such as hinges, latches, or seals, can cause clicking, squeaking, or rattling noises. Ensure that the door closes properly and seals tightly when the microwave is in operation.

Internal Components: Loose or damaged internal components, such as wiring, screws, or mounting brackets, can create rattling, buzzing, or vibrating noises during operation.

Metal Objects: If metallic objects or debris are accidentally placed inside the microwave, they can cause sparks, arcing, or popping noises, posing safety hazards and potentially damaging internal components.

Ventilation Grilles: Loose or improperly secured ventilation grilles or covers can vibrate or rattle during operation, producing unwanted noises.

How to repair:

Safety Precautions: Unplug the microwave from the power source before beginning any repairs. Use insulated tools and wear appropriate personal protective equipment (PPE) such as gloves and safety glasses. Be cautious of high-voltage components like the capacitor and magnetron, which can store dangerous electrical charges even when unplugged.

Diagnostic Steps: Identify the specific issue with the microwave. This could error codes displayed on the control panel, listening for unusual noises, or noting any abnormalities in its operation. Consult the microwave's manual for troubleshooting guidance and safety precautions specific to your model.

Visual Inspection: Carefully examine the internal and external components of the microwave for any signs of damage, wear, or loose connections. Look for burnt wires, charred components, or melted insulation, which may indicate electrical faults or overheating.

Testing Components: Use a multimeter to test the continuity of electrical components, such as the fuse, diode, capacitor, magnetron, and door switches. Check for proper voltage levels at various points in the circuit to ensure that power is reaching the appropriate components.

Repair or Replace: Based on your diagnostics, repair or replace any faulty components identified during testing. Follow the manufacturer's guidelines and use OEM (original equipment manufacturer) replacement parts whenever possible to ensure compatibility and safety. Common components that may need replacement include fuses, door switches, diodes, capacitors, magnetrons, and fan motors.

Reassembly and Testing: Reassemble the microwave carefully, ensuring that all components are properly connected and secured. Plug in the microwave and perform a test run to verify that the issue has been resolved and that the appliance functions correctly. Monitor the microwave during operation for any signs of abnormal behavior or recurring issues.

DISPLAY OR CONTROL PANEL ISSUES

Common issues with display or control panels in microwaves can include:

Unresponsiveness: The display or control panel may become unresponsive, making it difficult or impossible to operate the microwave. This could be due to a malfunctioning control board, faulty buttons, or issues with the display screen.

Erratic Behavior: The display or control panel may exhibit erratic behavior, such as flickering, flashing, or displaying incorrect information. This could indicate problems with the control board, loose connections, or electrical faults.

Partial Display or Dim Lighting: Parts of the display may be dimly lit or not illuminated at all, making it challenging to read the settings or information. This could be caused by a faulty display panel, defective backlighting, or issues with the control board.

Button Malfunction: Individual buttons on the control panel may stop working or become stuck, preventing the user from inputting commands or adjusting settings. This could be due to worn-out buttons, debris or spills obstructing the buttons, or problems with the underlying circuitry.

Error Codes: The display panel may show error codes indicating specific issues or malfunctions within the microwave. Common error codes include codes related to door issues, sensor problems, or electrical faults.

Interference or Noise: The display or control panel may experience interference or noise, such as buzzing or humming sounds, which can affect its functionality and readability. This could be caused by electrical interference, faulty components, or poor grounding.

Water Damage: Spills or moisture entering the control panel area can cause damage to the circuitry, leading to malfunctioning displays, unresponsive buttons, or short circuits.

Software Glitches: Occasionally, software glitches or firmware issues can cause problems with the display or control panel. Resetting the microwave or updating its firmware may resolve these issues.

Physical Damage: Physical damage to the display or control panel, such as cracks, scratches, or dents, can impair its functionality and readability. This could be due to accidental impacts, improper handling, or wear and tear over time.

Wiring or Connector Problems: Loose or damaged wiring connections or connectors between the control panel and the main circuit board can cause intermittent issues with the display or control functions.

Repairing issues with the display or control panel of a microwave can be complex and may require careful diagnosis and handling of electrical components. Here's a general guide on how to repair common issues with display or control panels:

Safety Precautions: Before starting any repairs, unplug the microwave from the power source to prevent the risk of electric shock. Use insulated tools and wear appropriate personal protective equipment (PPE), such as gloves and safety glasses, to protect yourself from potential hazards.

Diagnosis: Identify the specific issue with the display or control panel. Determine whether the problem is with the display screen, buttons, circuitry, or other components. Refer to the microwave's manual for troubleshooting guidance and error code meanings.

Visual Inspection: Carefully examine the display panel, buttons, and surrounding area for any signs of physical damage, such as cracks, scratches, or water damage. Check for loose connections, broken wires, or damaged components on the control board or behind the control panel.

Testing Components: Use a multimeter to test the continuity of electrical components, such as buttons, switches, wiring, and connectors. Test the display panel and backlighting to ensure they are receiving power and functioning correctly. Verify the operation of the control board and any integrated circuits responsible for processing user inputs and displaying information.

Repair or Replacement: Depending on the diagnosis, repair or replace the faulty components. Common repairs may include:

- Cleaning dirty or sticky buttons to restore functionality.

- Replacing damaged or worn-out buttons, switches, or membrane keypads.

- Repairing or replacing damaged wiring, connectors, or solder joints.

- Repairing minor physical damage to the display panel or control board.

- Replacing malfunctioning components such as the display screen, backlighting, or integrated circuits. Follow manufacturer's guidelines and use OEM (original equipment manufacturer) replacement parts whenever possible to ensure compatibility and safety.

Reassembly and Testing: Carefully reassemble the microwave, ensuring that all components are properly connected and secured. Plug in the microwave and perform a test run to verify that the issue has been resolved and that the display and control panel are functioning correctly. Test all buttons, settings, and display functions to ensure proper operation.

INSULATION PROBLEMS

Microwave Leakage: One of the primary functions of insulation in a microwave is to contain the electromagnetic radiation generated by the magnetron. If the insulation is damaged or compromised, it can result in microwave leakage, which poses a significant safety risk to users. Microwave leakage can cause exposure to harmful radiation and may lead to health problems.

Heat Loss: Insulation helps to maintain the internal temperature of the microwave cavity, allowing it to heat food efficiently. If the insulation is inadequate or damaged, it can lead to heat loss, which may result in longer cooking times, uneven heating, or decreased energy efficiency.

Condensation and Moisture Buildup: Insulation helps to prevent condensation and moisture buildup inside the microwave cavity. If the insulation is ineffective, moisture can accumulate, leading to rust, corrosion, and electrical problems. Additionally, excessive moisture can affect the performance of the microwave and increase the risk of malfunction.

Noise Reduction: Insulation also plays a role in reducing noise and vibrations generated during the operation of the microwave. If the insulation is insufficient or deteriorated, it may fail to dampen noise effectively, resulting in louder operation or rattling sounds.

Structural Integrity: Insulation contributes to the structural integrity of the microwave, providing support and stability to internal components. If the insulation is damaged or degraded, it can compromise the structural integrity of the microwave, leading to potential mechanical failures or safety hazards.

Common causes of insulation problems in microwaves include:

Physical Damage: Insulation can be damaged by impact, abrasion, or exposure to high temperatures. Common sources of physical damage include dropping the microwave, placing heavy objects on top of it, or using abrasive cleaning materials.

Moisture and Humidity: Exposure to moisture and humidity can degrade insulation over time, leading to reduced effectiveness and potential mold or mildew growth. This can occur if the microwave is installed in a damp environment or if spills and splashes are not cleaned up promptly.

Age and Wear: Over time, insulation materials may deteriorate due to age, use, and exposure to environmental factors. This can result in decreased insulation effectiveness and increased susceptibility to damage.

Repairing insulation problems in a microwave can be challenging and may require professional assistance, especially if the issue involves addressing microwave leakage or replacing internal insulation materials. However, here are some general steps you can take to address common insulation problems:

Safety Precautions: Before attempting any repairs, unplug the microwave from the power source to prevent the risk of electric shock. Wear appropriate personal protective equipment (PPE), such as gloves and safety glasses, to protect yourself from potential hazards.

Diagnosis: Identify the specific insulation problem with the microwave. This could include issues such as visible damage to the insulation material, evidence of moisture buildup, or concerns about microwave leakage. Perform a visual inspection of the microwave's interior and exterior to assess the condition of the insulation.

Repair Options: For minor insulation damage, such as small tears or cracks in the insulation material, you may be able to repair it using specialized insulation tape or sealant designed for microwave use. Follow the manufacturer's instructions for application and ensure that the repaired area is adequately sealed. If the insulation problem is due to moisture buildup, thoroughly clean and dry the affected areas. Ensure that the microwave is properly ventilated to prevent future moisture accumulation. If the microwave is leaking radiation due to damaged insulation, it's essential to address the issue promptly. However, repairing microwave leakage typically requires specialized equipment and expertise. In this case, it's best to contact a qualified technician or the manufacturer for assistance.

Preventive Measures: Once the insulation problem is addressed, take preventive measures to avoid future issues. This may include regular inspection and maintenance of the microwave, prompt cleaning of spills and splashes, and ensuring proper ventilation in the microwave's environment.

SPARKING OR ARCING

Sparking or arcing in a microwave is a serious issue that requires immediate attention, as it poses significant safety hazards and can lead to damage to the appliance. Common issues that cause sparking or arcing in a microwave include:

Metal Objects Inside the Microwave: Placing metal objects, such as aluminum foil, utensils with metal trim, or metal twist ties, inside the microwave can cause sparking or arcing. Metal reflects microwaves and creates a buildup of electrical charge, resulting in sparks.

Damaged Waveguide Cover: The waveguide cover is a protective panel that directs microwaves into the microwave cavity. If the waveguide cover is damaged, dirty, or has food splatter buildup, it can cause sparking or arcing as microwaves are reflected off the damaged surface.

Faulty or Damaged Stirrer Fan: Some microwaves are equipped with a stirrer fan that helps distribute microwaves evenly throughout the cavity. If the stirrer fan is damaged or malfunctioning, it can cause uneven distribution of microwaves and result in sparking or arcing.

Worn-out or Damaged Magnetron: The magnetron is the component that generates microwaves. If the magnetron is worn out, damaged, or malfunctioning, it can produce uneven or erratic microwave output, leading to sparking or arcing inside the microwave cavity.

Excessive Food Splatter or Grease: Food splatter or grease buildup inside the microwave cavity can accumulate over time and create conductive paths for electricity, resulting in sparking or arcing during operation.

Dirty or Damaged Interior Surfaces: Dirty or damaged interior surfaces, such as the walls, ceiling, or floor of the microwave cavity, can cause sparking or arcing. Food debris or spills that come into contact with these surfaces during cooking can ignite or create electrical arcs.

Overheating Components: Overheating components, such as the magnetron, diode, capacitor, or transformer, can cause sparking or arcing due to electrical faults or insulation breakdown.

Faulty High Voltage Diode or Capacitor: The high voltage diode and capacitor are components that help regulate the flow of electricity in the microwave's high voltage circuit. If either of these components is faulty or damaged, it can cause sparking or arcing.

Moisture or Condensation: Moisture or condensation inside the microwave cavity can create a conductive environment, leading to sparking or arcing. This can occur if the microwave is not adequately ventilated or if food with high moisture content is cooked without a cover.

Repairing sparking or arcing in a microwave requires careful attention to safety and adherence to proper procedures. Here's a general guide on how to repair this issue:

Safety Precautions: Before attempting any repairs, unplug the microwave from the power source to prevent the risk of electric shock. Wear appropriate personal protective equipment (PPE), such as gloves and safety glasses, to protect yourself from potential hazards.

Identify the Source of Sparking or Arcing: Inspect the interior of the microwave cavity for any visible signs of damage, such as burned spots, melted plastic, or metal objects. Remove any metal objects or debris that may be causing the sparking or arcing. Check the waveguide cover for signs of damage, such as cracks, burns, or food splatter buildup.

Clean the Microwave Cavity: Thoroughly clean the interior surfaces of the microwave cavity using a mild detergent or microwave-safe cleaner. Remove any food splatter, grease, or debris that may be contributing to the sparking or arcing. Pay special attention to the waveguide cover and ensure that it is clean and free of obstructions.

Inspect and Replace Damaged Components: If the waveguide cover is damaged or dirty, carefully remove it and replace it with a new one. Be sure to follow the manufacturer's instructions for installation. Inspect the stirrer fan, magnetron, high voltage diode, capacitor, and other internal components for signs of damage or malfunction. Replace any damaged or faulty components with new ones. Be sure to use replacement parts that are compatible with your microwave model.

Test the Microwave: Plug in the microwave and perform a test run to verify that the issue has been resolved. Run the microwave with a microwave-safe container of water for a short period to ensure that there is no sparking or arcing during operation.

Monitor the Microwave: Keep an eye on the microwave during operation to ensure that the problem does not recur. If you notice any signs of sparking or arcing, immediately stop using the microwave and reinspect for potential causes.

CONDENSATION BUILDUP

Condensation buildup inside a microwave can occur due to various factors, including improper usage, environmental conditions, and issues with the appliance itself. Here are some common issues associated with condensation buildup in microwaves:

Improper Covering of Food: When cooking or reheating food in the microwave, steam is generated, which can lead to condensation buildup inside the microwave cavity. If food is not properly covered with a microwave-safe lid or cover, steam can escape and condense on the interior surfaces of the microwave.

Inadequate Ventilation: Microwaves are equipped with ventilation systems to expel steam and moisture generated during cooking. If the ventilation system is blocked or not functioning correctly, moisture may accumulate inside the microwave cavity, leading to condensation buildup.

High Humidity Environment: Operating a microwave in a high humidity environment, such as a kitchen with poor ventilation or near a stove or dishwasher, can increase the likelihood of condensation buildup. Moisture from the air can enter the microwave cavity and condense on the cooler interior surfaces.

Damaged or Missing Seals: The door seals and gaskets on a microwave help to create an airtight seal when the door is closed, preventing moisture from escaping. If the seals are damaged, worn out, or missing, steam and moisture can leak out of the microwave cavity, leading to condensation buildup.

Cracked or Damaged Interior Surfaces: Cracks, chips, or damage to the interior surfaces of the microwave cavity can provide a pathway for moisture to enter and accumulate. Over time, condensation can collect in these damaged areas and lead to further deterioration.

Inadequate Insulation: Poor insulation within the microwave cavity can cause the interior surfaces to become cooler than the surrounding air, leading to condensation buildup. This can occur if the insulation material is damaged, degraded, or insufficient for maintaining temperature balance.

Cooking with Frozen Foods: Cooking frozen foods in the microwave without proper thawing or venting can release a significant amount of moisture into the cavity, increasing the risk of condensation buildup.

Extended Cooking Times: Prolonged cooking or heating cycles in the microwave can generate excess steam and moisture, especially if the food is not covered or vented properly. This can contribute to condensation buildup over time.

To address condensation buildup in a microwave, consider the following steps:

- Ensure that food is properly covered or vented during cooking to minimize steam escape.

- Clean and inspect the microwave's ventilation system, including the exhaust fan and filters, to ensure proper airflow.

- Check the door seals and gaskets for damage and replace them if necessary to maintain an airtight seal.

- Monitor the microwave's operation and avoid prolonged cooking times or cooking frozen foods without proper venting.

Lift Stations

PUMP FAILURE

Pump failure in lift stations can result from various factors, each of which can disrupt the efficient transport of wastewater and lead to potential environmental and public health hazards. Here are some common issues associated with pump failure in lift stations:

Motor Failure: The electric motor that drives the pump can fail due to various reasons, including overheating, bearing failure, electrical issues, or mechanical wear and tear. Motor failure can result in the pump becoming inoperable, causing sewage backups or overflows.

Clogging: Pump clogging is a prevalent issue in lift stations. It occurs when debris, solids, grease, or other materials accumulate in the pump intake or impeller, obstructing the flow of wastewater. Clogging can lead to decreased pump efficiency, increased energy consumption, and eventual pump failure if not addressed promptly.

Sediment and Sand Ingestion: Lift stations located in areas with sandy or sediment-rich wastewater may experience pump failure due to abrasion and wear on pump components. Sand and sediment can damage impellers, seals, and bearings, leading to reduced pump performance and eventual failure.

High Flow Rates: Periods of high flow rates, such as during heavy rainfall or peak usage times, can exceed the capacity of lift station pumps. Continuous operation under high flow conditions can strain the pumps, leading to overheating, motor failure, or premature wear of pump components.

Power Outages: Power outages or electrical failures can render lift station pumps inoperable, disrupting the flow of wastewater and potentially causing backups or overflows. Lift stations may be equipped with backup power sources such as generators or battery backups to mitigate the impact of power outages.

Float Switch Malfunction: Lift stations typically use float switches to control the operation of pumps based on water levels in the wet well. If the float switch malfunctions or becomes stuck, it can result in improper pump operation, leading to pump failure or overflows.

Corrosion and Erosion: Lift station pumps are exposed to corrosive wastewater and abrasive materials, which can cause corrosion and erosion of pump components over time. Corrosion can weaken pump housings, impellers, and seals, leading to leaks, reduced efficiency, and eventual pump failure.

Lack of Maintenance: Inadequate maintenance practices, such as infrequent inspections, lubrication, or cleaning, can contribute to pump failure in lift stations. Neglected pumps are more prone to issues such as clogging, overheating, and mechanical failures, leading to costly repairs and downtime.

Incorrect Pump Sizing: Improperly sized pumps may be unable to handle the wastewater flow rates or head pressures encountered in the lift station, leading to frequent breakdowns or premature failure. Proper pump selection and sizing based on hydraulic calculations are essential to ensure reliable operation.

> **Repairing pump failure in lift stations requires a systematic approach to identify and address the root cause of the problem. Here's a general guide on how to repair pump failure in lift stations:**

Safety Precautions: Before beginning any repairs, ensure that the lift station is properly isolated from electrical power sources to prevent the risk of electric shock. Use appropriate personal protective equipment (PPE), including gloves, safety glasses, and protective clothing, to safeguard against potential hazards.

Diagnosis: Identify the specific cause of pump failure by conducting a thorough inspection of the lift station and pump system. Check for visible signs of damage, such as clogging, corrosion, leaks, or mechanical wear and tear. Use diagnostic tools, such as pressure gauges, flow meters, or vibration analyzers, to assess pump performance and identify abnormalities.

Clearing Clogs and Blockages: If pump failure is due to clogging or blockages, remove debris, solids, or obstructions from the pump intake, impeller, or discharge piping. Use appropriate tools, such as sewer rods, jetting equipment, or mechanical augers, to clear clogs and restore proper flow.

Repair or Replace Pump Components: Repair or replace damaged pump components, such as impellers, seals, bearings, or motor parts, as needed. Follow the manufacturer's guidelines and specifications for pump repair and replacement procedures. Use OEM (original equipment manufacturer) replacement parts to ensure compatibility and performance.

Electrical Troubleshooting: Check electrical connections, wiring, and controls to identify and address any issues related to power supply, motor operation, or control system malfunctions. Test electrical components, such as capacitors, relays, and contactors, for proper functionality and continuity. Ensure that float switches, level sensors, and control panels are operating correctly and adjusting pump operation as intended.

Preventive Maintenance: Implement a preventive maintenance program for the lift station and pump system to prevent future failures. Schedule regular inspections, cleaning, lubrication, and testing of pump components and system controls. Keep accurate records of maintenance activities, repairs, and equipment performance for tracking and analysis.

Testing and Commissioning: After completing repairs, test the pump system under normal operating conditions to verify proper functionality and performance. Monitor pump operation, flow rates, pressure levels, and system alarms to ensure that the repair has effectively resolved the issue. Conduct commissioning tests, such as pump run tests, flow tests, and alarm tests, to validate the integrity and reliability of the lift station system.

FLOAT SWITCH MALFUNCTION

Float switches are crucial components in lift stations that control the operation of pumps based on water levels in the wet well. When float switch malfunctions occur, it can lead to improper pump operation, potentially causing pump failures, system overflows, and other operational issues. Common issues associated with float switch malfunction in lift stations include:

Stuck Float Switch: Float switches may become stuck in the "on" or "off" position due to debris accumulation, mechanical obstructions, or physical damage. A stuck float switch can prevent the pump from activating or deactivating as intended, resulting in improper water level control and potential system malfunctions.

Fouled Float Switch: Accumulation of debris, grease, or biological growth on the surface of float switches can impair their buoyancy and hinder proper operation. A fouled float switch may fail to rise or fall with changes in water level, leading to inaccurate level sensing and unreliable pump control.

Corrosion or Rust: Exposure to corrosive wastewater or environmental conditions can cause corrosion or rusting of float switch components, such as floats, rods, or electrical contacts. Corrosion can interfere with the electrical conductivity of the switch, resulting in erratic or intermittent operation.

Electrical Issues: Faulty wiring, loose connections, or damaged electrical components associated with float switches can cause electrical malfunctions and lead to float switch failure. Electrical problems may result in loss of signal transmission, inaccurate level sensing, or inability to activate/deactivate the pump.

Mechanical Wear and Tear: Over time, mechanical components of float switches, such as pivot points, hinges, or seals, may experience wear and deterioration. Mechanical wear can compromise the reliability and precision of float switch operation, increasing the risk of malfunction.

Environmental Factors: Environmental factors such as temperature fluctuations, humidity, exposure to UV radiation, and exposure to chemicals can affect the performance and longevity of float switches. Extreme environmental conditions may accelerate degradation and lead to premature failure.

Improper Installation or Adjustment: Incorrect installation, calibration, or adjustment of float switches can result in improper positioning or sensitivity, leading to unreliable level sensing and pump control. Float switches must be installed according to the manufacturer's specifications and properly calibrated for optimal performance.

Float Interference: Interference from floating debris, floating solids, or other objects in the wet well can obstruct the movement of float switches and interfere with their operation. Float interference can prevent accurate level sensing and compromise pump control.

Age and Deterioration: Like any mechanical or electrical component, float switches are subject to wear, aging, and deterioration over time. As float switches age, their performance may degrade, increasing the likelihood of malfunction and failure.

Repairing float switch malfunctions in lift stations involves a systematic approach to identify, troubleshoot, and resolve issues affecting the operation of the float switches. Here's a general guide on how to repair float switch malfunctions in lift stations:

Safety Precautions: Before beginning any repairs, ensure that the lift station is properly isolated from electrical power sources to prevent the risk of electric shock. Use appropriate personal protective equipment (PPE), including gloves, safety glasses, and protective clothing, to safeguard against potential hazards.

Diagnosis: Identify the specific symptoms and behaviors of the float switch malfunction, such as failure to activate/deactivate the pump, erratic operation, or inaccurate level sensing. Inspect the float switch assembly and associated components for visible signs of damage, corrosion, fouling, or mechanical wear.

Cleaning and Maintenance: Clean the float switch assembly and surrounding area to remove debris, grease, or biological growth that may be obstructing its operation. Check for fouling or buildup on the float switch floats, rods, or electrical contacts, and clean or replace them as needed. Inspect mechanical components, such as pivot points, hinges, and seals, for wear and deterioration, and lubricate or replace them as necessary.

Electrical Troubleshooting: Test the electrical connections, wiring, and terminals associated with the float switch for continuity, loose connections, or damage. Use a multimeter or continuity tester to check for electrical faults, shorts, or open circuits in the float switch circuitry. Verify the integrity of electrical components, such as relays, switches, and control panels, and repair or replace any faulty components.

Calibration and Adjustment: Calibrate the float switch according to the manufacturer's specifications to ensure accurate level sensing and reliable pump control. Adjust the positioning, sensitivity, or activation/deactivation points of the float switch as needed to optimize its performance. Verify that the float switch responds correctly to changes in water level and activates/deactivates the pump as intended.

Replacement of Components: If the float switch or its components are damaged beyond repair or cannot be restored to proper functionality, replace them with new ones. Use OEM (original equipment manufacturer) replacement parts to ensure compatibility, performance, and reliability.

Testing and Verification: Test the repaired or replaced float switch under normal operating conditions to verify proper functionality and performance. Monitor the float switch operation, pump activation/deactivation, and water level control to ensure that the repair has effectively resolved the issue.

Preventive Maintenance: Implement a preventive maintenance program for the lift station and float switch system to prevent future malfunctions. Schedule regular inspections, cleaning, lubrication, and testing of float switch components and associated equipment. Keep accurate records of maintenance activities, repairs, and equipment performance for tracking and analysis.

CLOGGING AND BLOCKAGES

Clogging and blockages are common issues in lift stations that can disrupt the flow of wastewater, leading to pump failures, system backups, and potential environmental hazards. Here are some common issues associated with clogging and blockages in lift stations:

Debris Accumulation: Lift stations may accumulate debris such as leaves, trash, plastic bags, or sanitary products, which can obstruct the flow of wastewater and lead to clogging in pump intakes, discharge pipes, or valves.

Grease Buildup: Fats, oils, and grease (FOG) from kitchen waste can solidify and adhere to the walls of pipes and pump components, causing blockages and restricting the flow of wastewater.

Sediment Deposition: Sediment, sand, grit, or other solids present in wastewater can settle and accumulate in the wet well or pump sump, leading to sedimentation and eventual blockages in pump intakes or impellers.

Tree Root Intrusion: Tree roots can infiltrate sewer lines and penetrate lift station wet wells or discharge pipes, causing obstructions and blockages that impede the flow of wastewater.

Ragging and Rags: Fibrous materials such as wipes, rags, towels, or textiles can become entangled in pump impellers or mechanical components, causing pump blockages and reducing pump efficiency.

Inflow and Infiltration (I&I): Excessive inflow and infiltration of stormwater or groundwater into the sewer system can overwhelm lift stations, leading to the accumulation of debris, sediment, or floatable materials that contribute to clogging and blockages.

Foreign Objects: Accidental or intentional disposal of foreign objects such as toys, tools, or construction debris into sewer lines or lift station wet wells can cause blockages and damage to pump components.

Collapsed or Damaged Pipes: Structural defects, cracks, or collapses in sewer lines or lift station discharge pipes can obstruct the flow of wastewater, leading to backups and blockages.

Corrosion and Scaling: Corrosion of metal pipes or scaling of interior surfaces due to chemical reactions or mineral deposits can contribute to the accumulation of debris and sediment, leading to blockages and reduced flow capacity.

Improper Pump Sizing: Inadequately sized pumps may be unable to handle the flow rates or head pressures encountered in the lift station, leading to frequent blockages and pump failures.

Addressing clogging and blockages in lift stations typically involves a combination of preventive measures, maintenance practices, and corrective actions. Here are some strategies for addressing clogging and blockages:

Regular Cleaning: Implement a routine cleaning schedule to remove debris, sediment, and grease buildup from lift station wet wells, pump intakes, and discharge pipes.

Screening and Grinders: Install screens, grinders, or comminutors to capture and shred solid materials before they enter the lift station pumps, reducing the risk of blockages and pump damage.

Flushable Wipes and Grease Management: Educate the public about the proper disposal of flushable wipes and grease to prevent them from entering the sewer system and causing blockages in lift stations.

Root Control: Use root control treatments or barriers to prevent tree roots from infiltrating sewer lines and lift station wet wells, reducing the risk of blockages and structural damage.

Pump Maintenance: Conduct regular inspections, lubrication, and cleaning of lift station pumps to prevent pump blockages and ensure reliable operation.

I&I Reduction: Implement measures to reduce inflow and infiltration into the sewer system, such as repairing leaks, sealing manholes, and installing backflow prevention devices.

Pipe Inspection and Repair: Inspect sewer lines and lift station discharge pipes regularly for signs of damage, corrosion, or blockages, and repair or replace damaged sections as needed.

Proactive Monitoring: Use telemetry systems, alarms, and level sensors to monitor lift station performance and detect signs of clogging or blockages early, allowing for timely intervention and corrective action.

ELECTRICAL PROBLEMS

Electrical problems in lift stations can disrupt the operation of pumps, control systems, and other electrical components, leading to system failures, pump malfunctions, and potential safety hazards. Here are some common electrical problems associated with lift stations:

Power Supply Issues: Problems with the main power supply, including power surges, voltage fluctuations, or intermittent power outages, can affect the operation of lift station equipment and control systems.

Motor Failures: Electric motors used in lift station pumps may experience failures due to issues such as overheating, bearing wear, electrical faults, or mechanical failures. Motor failures can result in pump malfunctions and system downtime.

Control Panel Malfunctions: Lift station control panels, including motor starters, relays, contactors, and PLCs (programmable logic controllers), may experience malfunctions due to electrical faults, component failures, or programming errors. Control panel malfunctions can disrupt the automatic operation of pumps and other equipment.

Float Switch Problems: Float switches are critical components in lift stations that control the operation of pumps based on water levels in the wet well. Electrical issues such as wiring faults, loose connections, or switch failures can cause float switch malfunctions, leading to improper pump control and potential system overflows.

Grounding Problems: Inadequate grounding or improper grounding of electrical equipment in lift stations can increase the risk of electrical hazards such as electric shock, equipment damage, or stray currents. Grounding problems may occur due to corroded grounding connections, poor soil conductivity, or improper installation.

Electrical Shorts and Open Circuits: Short circuits, open circuits, or wiring faults in electrical circuits can disrupt the flow of electricity and cause equipment malfunctions or failures. Electrical shorts may result from damaged insulation, loose connections, or exposure to moisture.

Corrosion and Moisture Damage: Lift station electrical components are exposed to harsh environmental conditions, including moisture, humidity, and corrosive gases. Corrosion and moisture damage can degrade electrical insulation, corrode metal contacts, and cause electrical failures in equipment such as switches, relays, and wiring.

Overloading and Overheating: Overloading of electrical circuits or components beyond their rated capacity can lead to overheating, insulation breakdown, and component failures. Overloading may occur due to excessive current draw, inadequate wire sizing, or insufficient cooling.

Aging Infrastructure: Aging lift station infrastructure, including electrical wiring, conduit, and equipment, may deteriorate over time due to factors such as corrosion, thermal cycling, and mechanical stress. Aging infrastructure can increase the risk of electrical problems and equipment failures.

Lack of Maintenance: Inadequate maintenance practices, such as infrequent inspections, cleaning, or testing of electrical equipment, can contribute to the accumulation of electrical problems and increase the risk of equipment failures. Regular maintenance is essential to identify and address electrical issues before they escalate.

Repairing electrical problems in lift stations requires careful attention to safety and adherence to proper procedures. Here's a general guide on how to repair common electrical problems in lift stations:

Safety Precautions: Before beginning any repairs, ensure that the lift station is properly isolated from electrical power sources to prevent the risk of electric shock. Use appropriate personal protective equipment (PPE), including gloves, safety glasses, and insulated tools, to safeguard against potential hazards.

Diagnosis: Identify the specific symptoms and behaviors of the electrical problem, such as equipment malfunctions, power fluctuations, or circuit failures. Inspect electrical components, wiring, and connections for visible signs of damage, corrosion, loose connections, or overheating.

Isolation and Testing: Isolate the affected electrical circuit or component from the power source by switching off the circuit breaker or disconnecting power from the main panel. Use a multimeter or voltage tester to test for voltage presence, continuity, and proper operation of electrical circuits, switches, relays, and components.

Repair or Replace Components: Repair or replace damaged electrical components, such as switches, relays, contactors, fuses, circuit breakers, or wiring, as needed. Follow the manufacturer's guidelines and specifications for component repair and replacement procedures. Use OEM (original equipment manufacturer) replacement parts to ensure compatibility, performance, and safety.

Wire and Connection Repair: Repair or replace damaged or corroded electrical wiring, connectors, terminals, and junction boxes as needed. Use appropriate wire splices, connectors, and terminal blocks rated for the voltage and current requirements of the lift station equipment.

Grounding and Bonding: Check and repair grounding and bonding connections to ensure proper electrical grounding of lift station equipment and components. Ensure that grounding conductors are securely connected to grounding electrodes and bonded to metal equipment and structures.

Insulation and Enclosure Integrity: Inspect electrical insulation, enclosures, and protective covers for damage, deterioration, or exposure to moisture. Repair or replace damaged insulation, seals, gaskets, and enclosure components to maintain electrical safety and protection from environmental hazards.

Testing and Verification: After completing repairs, test the electrical circuit or component under normal operating conditions to verify proper functionality and performance. Monitor voltage levels, current draw, and equipment operation to ensure that the repair has effectively resolved the electrical problem.

Preventive Maintenance: Implement a preventive maintenance program for the lift station electrical system to prevent future problems. Schedule regular inspections, cleaning, testing, and calibration of electrical components and circuits to identify and address issues before they escalate.

Documentation and Recordkeeping: Keep detailed records of electrical repairs, component replacements, testing results, and maintenance activities for future reference and analysis. Document any modifications, upgrades, or improvements made to the lift station electrical system to maintain an accurate maintenance history.

FLOAT AND VALVE ISSUES

Floats and valves play crucial roles in the operation of lift stations, controlling the levels of wastewater within the wet well and regulating the activation of pumps. Several common issues can arise with floats and valves in lift stations, potentially leading to operational problems and system failures. Here are some common float and valve issues:

FLOAT SWITCH MALFUNCTIONS:

Stuck Floats: Floats can become stuck in the "on" or "off" position due to debris accumulation, mechanical obstructions, or damage, resulting in improper pump activation or deactivation.

Fouled Floats: Accumulation of debris, grease, or biological growth on float surfaces can affect buoyancy and hinder proper operation, leading to inaccurate level sensing and unreliable pump control.

Electrical Failures: Wiring faults, loose connections, or switch failures can cause electrical malfunctions in float switches, resulting in erratic operation or complete failure to signal pump activation or deactivation.

CHECK VALVE ISSUES:

Clogging: Check valves can become clogged with debris, sediment, or grease, impeding the flow of wastewater and preventing proper pump operation.

Leaks: Damaged or deteriorated check valve seals can cause leaks, allowing wastewater to backflow into the wet well and increasing the risk of overflows or system backups.

Valve Sticking: Check valves may stick in the closed position due to debris buildup, corrosion, or mechanical issues, preventing proper flow regulation and reducing pump efficiency.

GATE VALVE PROBLEMS:

Sticking or Binding: Gate valves may become stuck or bind due to debris accumulation, corrosion, or improper lubrication, inhibiting their ability to open or close fully.

Leaks: Damaged seals or valve seats can cause leaks in gate valves, resulting in loss of pressure, reduced pump performance, or system inefficiencies.

Corrosion: Exposure to corrosive wastewater or environmental conditions can cause corrosion of gate valve components, leading to deterioration and eventual failure.

AIR RELEASE VALVE ISSUES:

Air Locks: Air release valves can become air-locked, preventing the release of trapped air from the system, and causing pump cavitation, reduced flow rates, or pump damage.

Malfunctioning Floats: Float-operated air release valves may malfunction due to debris accumulation, mechanical issues, or improper adjustment, affecting their ability to vent air from the system effectively.

FLOAT VALVE PROBLEMS:

Sticking or Binding: Float valves can stick or bind due to debris accumulation, corrosion, or mechanical issues, preventing proper regulation of water levels and pump activation.

Float Damage: Floats can become damaged or deteriorated over time, affecting their buoyancy and accuracy in controlling water levels within the wet well.

Mechanical Failures: Float valve assemblies may experience mechanical failures, such as broken linkages, worn seals, or damaged components, leading to improper operation and potential system failures.

PIPING AND CONNECTION ISSUES:

Blockages: Blockages in piping or connections leading to floats or valves can restrict flow and interfere with proper operation, resulting in inaccurate level sensing or valve control.

Leaks: Loose or damaged connections, fittings, or seals in piping systems can cause leaks, reducing system efficiency and increasing the risk of water infiltration or system damage.

> **Repairing float and valve issues in lift stations requires careful inspection, troubleshooting, and corrective action. Here's a general guide on how to repair common float and valve problems:**

Safety Precautions: Before beginning any repairs, ensure that the lift station is properly isolated from electrical power sources and that appropriate safety measures are in place to prevent accidents or injuries. Use personal protective equipment (PPE), including gloves, safety glasses, and protective clothing, to protect against hazards associated with working in a wastewater environment.

Diagnosis: Identify the specific symptoms and behaviors of the float or valve problem, such as improper pump operation, erratic water level fluctuations, or leaks. Inspect float switches, float valves, check valves, gate valves, and air release valves for visible signs of damage, corrosion, or mechanical issues.

Cleaning and Maintenance: Clean float surfaces, valve seats, seals, and mechanical components to remove debris, grease, or fouling that may be affecting operation. Lubricate moving parts and mechanisms to ensure smooth operation and reduce the risk of binding or sticking.

Repair or Replace Components: Repair or replace damaged or worn float switches, float valves, check valves, gate valves, or air release valves as needed. Follow the manufacturer's guidelines and specifications for component repair and replacement procedures. Use OEM (original equipment manufacturer) replacement parts to ensure compatibility and performance.

Electrical Troubleshooting: Test float switch circuits for continuity, proper wiring connections, and voltage levels using a multimeter or voltage tester. Check for loose connections, damaged wiring, or electrical faults in float switch assemblies and associated control panels. Repair or replace faulty electrical components, such as relays, contactors, or wiring, as needed. Adjustment and Calibration: Adjust float switch settings, positioning, or sensitivity to ensure accurate level sensing and reliable pump control. Calibrate float valves or control systems to maintain desired water levels within the wet well and prevent overflows or pump cycling.

Testing and Verification: After completing repairs, test the float or valve system under normal operating conditions to verify proper functionality and performance. Monitor water levels, pump operation, and system pressures to ensure that the repair has effectively resolved the problem.

Preventive Maintenance: Implement a preventive maintenance program for float and valve systems to prevent future issues. Schedule regular inspections, cleaning, lubrication, and testing of float switches, float valves, check valves, gate valves, and air release valves. Keep accurate records of maintenance activities, repairs, and equipment performance for tracking and analysis.

Documentation and Training: Document repair procedures, component replacements, and testing results for future reference and analysis. Provide training to operators and maintenance personnel on proper operation, maintenance, and troubleshooting of float and valve systems.

INFLOW LINE AND INFILTRATION

Inflow and infiltration (I&I) are common issues affecting lift stations, particularly in older or deteriorating sewer systems. These problems occur when excessive amounts of stormwater, groundwater, or other sources of clean water enter the sewer system, overwhelming the lift station's capacity and leading to various operational challenges. Common issues associated with lift station inflow and infiltration include:

Excessive Stormwater Inflow: During heavy rainfall events, stormwater can enter the sewer system through cracks, defects, or improperly sealed manholes, overwhelming lift station capacities and causing backups or overflows. In areas with combined sewer systems, stormwater runoff can inundate sanitary sewers, leading to surges in flow rates and hydraulic overloading of lift stations.

Groundwater Infiltration: Groundwater can infiltrate sewer pipes through cracks, joints, or deteriorated pipe materials, contributing to sustained flow rates and hydraulic loading on lift station pumps. Infiltrating groundwater may contain dissolved solids, sediment, or debris that can exacerbate pump wear and increase maintenance requirements.

Leaky Manholes and Structures: Leaky manholes, inspection chambers, or pump station structures can allow stormwater or groundwater to enter the sewer system, adding to inflow and infiltration volumes during wet weather conditions. Cracked or deteriorated manhole covers, frames, or seals can compromise the integrity of sewer infrastructure, leading to increased I&I rates.

Pipe Defects and Damage: Aging or deteriorating sewer pipes can develop defects such as cracks, fractures, or joint separations, providing pathways for stormwater or groundwater infiltration into the sewer system. Structural damage from root intrusion, corrosion, or soil settlement can exacerbate pipe defects and contribute to higher I&I rates over time.

High-Flow Sources: Direct connections of roof drains, downspouts, sump pumps, or foundation drains to the sanitary sewer system can introduce significant volumes of clean water into lift stations during rain events, increasing flow rates and operational stresses. Illicit connections, such as illegal tie-ins from surface water sources or foundation drains, can further exacerbate I&I issues and overload lift station capacities.

Hydraulic Overloading: Excessive inflow and infiltration can overwhelm lift station capacities, causing pumps to operate at or near maximum flow rates and increasing the risk of system backups, overflows, or pump failures. Hydraulic overloading can strain pump motors, increase energy consumption, and reduce the efficiency and effectiveness of lift station operations.

Impact on Treatment Facilities: Inflow and infiltration contribute to higher volumes of wastewater entering treatment facilities, potentially exceeding design capacities and treatment capabilities. Increased hydraulic and organic loading from I&I can disrupt treatment processes, compromise effluent quality, and lead to compliance issues with regulatory discharge standards.

Repairing inflow and infiltration (I&I) issues in lift stations requires a comprehensive approach that involves identifying sources of clean water ingress, assessing the extent of the problem, and implementing targeted mitigation strategies. Here's a general guide on how to repair common I&I issues in lift stations:

Source Identification: Conduct a thorough inspection of the lift station and surrounding sewer infrastructure to identify sources of inflow and infiltration, including leaky manholes, deteriorated pipes, faulty connections, and surface water sources. Use dye testing, smoke testing, CCTV inspection, flow monitoring, and other diagnostic techniques to pinpoint specific sources and pathways of clean water ingress.

Assessment and Prioritization: Evaluate the severity and impact of identified I&I sources on lift station performance, hydraulic capacity, and system reliability. Prioritize repair efforts based on factors such as flow rates, infiltration volumes, proximity to critical infrastructure, and cost-effectiveness of mitigation measures.

Repair and Rehabilitation Techniques: Implement targeted repair and rehabilitation techniques to address identified sources of inflow and infiltration, including:

Manhole Rehabilitation: Seal leaky manholes with gaskets, mortar, or epoxy coatings to prevent stormwater or groundwater ingress.

Pipe Repair: Repair or replace deteriorated sewer pipes using methods such as cured-in-place pipe (CIPP) lining, slip lining, or pipe bursting to restore structural integrity and reduce infiltration.

Joint Sealing: Seal pipe joints and connections using compression seals, gaskets, or chemical grouting to prevent water infiltration.

Service Lateral Inspection and Repair: Inspect and repair service laterals connecting individual properties to the sewer system to eliminate sources of clean water inflow.

Sump Pump Disconnection: Redirect roof drains, downspouts, foundation drains, and sump pump discharges away from the sanitary sewer system to reduce hydraulic loading on lift stations.

Stormwater Management: Implement green infrastructure practices, such as rain gardens, bioswales, and permeable pavement, to capture and infiltrate stormwater runoff before it reaches the sewer system.

Hydraulic Modeling and Analysis: Conduct hydraulic modeling and analysis to assess the effectiveness of I&I mitigation measures in reducing flow rates, minimizing pump cycling, and optimizing lift station performance. Use modeling results to refine repair strategies, prioritize future investments, and evaluate the long-term sustainability of I&I reduction efforts.

Monitoring and Maintenance: Implement a monitoring program to track inflow and infiltration rates, pump performance, and system response to repair interventions over time. Conduct regular maintenance activities, such as cleaning, inspection, and testing, to ensure the continued effectiveness of repair measures and identify any emerging issues.

Community Engagement and Education: Engage stakeholders, including property owners, businesses, and community groups, in efforts to reduce inflow and infiltration through public education campaigns, outreach events, and incentives for compliance with best management practices.

Regulatory Compliance and Reporting: Ensure compliance with regulatory requirements related to inflow and infiltration mitigation, wastewater discharge, and environmental protection. Maintain accurate records of repair activities, inspection results, and monitoring data for reporting purposes and regulatory compliance.

CORROSION AND WEAR

Corrosion and wear are common issues affecting lift stations due to the harsh operating environment and the presence of corrosive substances in wastewater. These problems can compromise the structural integrity, reliability, and performance of lift station components over time. Here are some common corrosion and wear issues encountered in lift stations:

Corrosion of Metal Components: Metal components such as pump casings, impellers, piping, valves, and structural supports are susceptible to corrosion from exposure to corrosive wastewater, gases, and chemicals. Corrosion can lead to metal loss, pitting, cracking, and structural weakening, compromising the integrity and longevity of lift station infrastructure.

Concrete Deterioration: Concrete structures, including wet wells, pump chambers, manholes, and pipe encasements, can deteriorate due to chemical attack, freeze-thaw cycles, and abrasion from solids in wastewater. Corrosion of embedded reinforcement steel (rebar) can lead to spalling, cracking, and delamination of concrete surfaces, reducing structural stability and increasing maintenance requirements.

Abrasion from Solids: Abrasive solids, such as sand, grit, gravel, and debris, can cause wear and erosion of pump impellers, volutes, wear rings, and piping surfaces. Abrasion can result in loss of pump efficiency, reduced flow rates, increased energy consumption, and premature failure of pump components.

Biofouling and Biological Growth: Microbial activity, algae, and biofilm formation within lift station components can contribute to corrosion, fouling, and deterioration of surfaces. Biofouling can reduce hydraulic efficiency, restrict flow passages, and increase maintenance requirements for cleaning and disinfection.

Electrochemical Corrosion: Electrochemical reactions, such as galvanic corrosion, can occur when dissimilar metals come into contact in the presence of an electrolyte, such as wastewater. Galvanic corrosion can accelerate metal deterioration and lead to premature failure of components, particularly in mixed-metal systems.

Chemical Attack: Chemicals present in wastewater, such as hydrogen sulfide (H_2S), sulfuric acid (H_2SO_4), chlorides, and acids, can cause chemical corrosion and degradation of metallic and concrete surfaces. A chemical attack can result in corrosion-induced cracking, loss of material strength, and permeability changes in concrete structures.

Cavitation Damage: Cavitation occurs when rapid changes in fluid pressure cause the formation and collapse of vapor bubbles, leading to localized erosion and pitting on pump impellers, wear rings, and valve seats. Cavitation damage can reduce pump performance, increase noise levels, and accelerate wear on critical components.

External Environmental Factors: Exposure to environmental factors such as temperature fluctuations, humidity, UV radiation, and airborne pollutants can accelerate corrosion and wear of lift station components. Protective coatings, paints, and corrosion-resistant materials may degrade over time, requiring periodic maintenance and recoating.

Repairing corrosion and wear issues in lift stations involves several steps to address the damage, restore structural integrity, and prevent further deterioration. Here's a general guide on how to repair common corrosion and wear issues in lift stations:

Assessment and Inspection: Conduct a comprehensive assessment and inspection of lift station components to identify areas affected by corrosion, wear, or deterioration. Use visual inspection, non-destructive testing (NDT) techniques, such as ultrasonic testing or magnetic particle inspection, and material analysis to evaluate the extent of damage and identify underlying causes.

Surface Preparation: Clean and prepare surfaces to be repaired by removing loose debris, rust, scale, corrosion products, and existing coatings. Use abrasive blasting, power washing, wire brushing, or chemical cleaning methods to achieve clean and properly prepared surfaces for repair.

Repair Techniques: Select appropriate repair techniques and materials based on the type and severity of corrosion and wear damage encountered. Common repair techniques for corrosion and wear issues in lift stations include:

Patching and Resurfacing: Apply polymer-modified mortars, epoxy coatings, or repair mortars to fill cracks, voids, and surface defects on concrete structures.

Metal Coatings: Apply corrosion-resistant coatings, such as epoxy, polyurethane, or zinc-rich paints, to protect metal surfaces from further corrosion and wear.

Welding and Metal Fabrication: Repair or replace corroded or worn metal components using welding techniques, metal overlays, or fabrication of replacement parts.

Linings and Liners: Install protective linings, liners, or coatings on pump casings, impellers, and piping systems to prevent abrasion, erosion, and chemical attack.

Cathodic Protection: Implement cathodic protection systems, sacrificial anodes, or impressed current systems to mitigate corrosion of metal components through electrochemical methods.

Quality Assurance: Ensure that repair materials, techniques, and procedures comply with industry standards, manufacturer specifications, and regulatory requirements. Conduct quality assurance checks, including adhesion tests, thickness measurements, and surface inspections, to verify the effectiveness and durability of the repair work.

Preventive Maintenance: Implement preventive maintenance measures to extend the service life of repaired components and minimize the risk of future corrosion and wear issues. Schedule regular inspections, cleanings, and recoating of repaired surfaces, and monitor cathodic protection systems and corrosion rates.

Documentation and Recordkeeping: Maintain detailed records of repair activities, including inspection reports, repair procedures, material specifications, and quality assurance documentation. Document any changes to lift station infrastructure, including repairs, upgrades, or modifications, for future reference and analysis.

Training and Education: Provide training to maintenance personnel on proper repair techniques, safety protocols, and preventive maintenance practices for addressing corrosion and wear issues in lift stations. Educate operators and staff on the importance of proactive maintenance, monitoring, and corrosion management strategies to ensure the long-term integrity and reliability of lift station infrastructure.

ALARM FAILURE

Alarm failures in lift stations can lead to serious operational issues, including equipment damage, system failures, and environmental hazards. Identifying and addressing alarm failures promptly is essential for maintaining the reliability and effectiveness of lift station operations. Here are some common alarm failures encountered in lift stations:

Power Failure Alarm Failure: Power failure alarms are designed to activate when there is a loss of electrical power to the lift station. Common causes of power failure alarm failures include malfunctioning power sensors, faulty wiring, or improper settings in the alarm system. Failure of the power failure alarm can result in pump shutdowns, system backups, and potential overflows during power outages.

High Water Level Alarm Failure: High water level alarms are intended to activate when the wastewater level in the wet well exceeds a predetermined threshold. Issues such as sensor malfunctions, wiring faults, or improper calibration can lead to high water level alarm failures. Failure of the high water level alarm can result in overflows, equipment damage, and environmental contamination due to inadequate response to rising water levels.

Pump Failure Alarm Failure: Pump failure alarms are designed to alert operators when lift station pumps fail to operate or experience abnormal conditions. Causes of pump failure alarm failures may include sensor defects, electrical faults, or communication errors within the alarm system. Failure to detect pump failures can lead to untreated wastewater discharges, system backups, and potential environmental violations.

Communication Failure Alarm Failure: Lift stations may be equipped with communication systems to relay alarm notifications to operators or remote monitoring centers. Communication failure alarms can fail to activate due to network issues, signal interference, or equipment malfunctions. Failure of communication failure alarms can result in delayed response times, missed alarms, and reduced situational awareness for operators.

Battery Backup Failure: Lift station alarm systems often incorporate battery backup systems to ensure continued operation during power outages. Battery backup failure alarms may occur due to battery defects, charging system failures, or inadequate backup power capacity. Failure of battery backup systems can compromise alarm functionality during power outages, leading to unmonitored lift station conditions and potential risks.

Sensor Calibration and Maintenance Issues: Incorrect sensor calibration, lack of routine maintenance, or sensor drift over time can lead to inaccurate alarm readings or false alarms. Regular calibration, inspection, and cleaning of sensors are essential for ensuring the reliability and accuracy of alarm systems.

Software or Control System Failures: Malfunctions or software glitches in the control system or alarm monitoring software can lead to alarm failures. Software updates, system diagnostics, and periodic testing are necessary to identify and rectify potential software-related issues.

Repairing alarm failures in lift stations requires a systematic approach to identify and address the underlying causes of the failures. Here's a step-by-step guide on how to repair common alarm failures in lift stations:

Diagnosis and Troubleshooting: Begin by conducting a thorough diagnosis of the alarm system to identify the specific alarm failure(s) and the root cause(s) of the problem. Inspect alarm components such as sensors, wiring, control panels, communication devices, and backup power systems for signs of damage, malfunction, or misconfiguration. Use diagnostic tools, multimeters, and testing equipment to verify the functionality of alarm system components and identify any faulty or defective parts.

Address Power Supply Issues: If the power failure alarm is not functioning properly, check the power supply to the lift station and ensure that electrical connections are secure and that voltage levels are within the specified range. Replace malfunctioning power sensors, fuses, circuit breakers, or backup batteries as needed to restore power supply monitoring capabilities.

Calibrate Sensors and Control Systems: Calibrate high water level sensors, pump failure sensors, and other alarm system components according to manufacturer specifications and industry standards. Verify the accuracy of sensor readings and adjust calibration settings as necessary to ensure reliable alarm activation and response.

Inspect and Repair Communication Systems: Check communication devices, telemetry equipment, and network connections to identify any issues affecting alarm transmission and reception. Troubleshoot communication failures, signal interference, or network connectivity problems and implement corrective actions, such as reconfiguring settings or replacing faulty hardware.

Replace Faulty Components: Replace malfunctioning or damaged alarm system components, such as sensors, relays, control panels, or communication modules, with new or refurbished parts. Ensure that replacement components are compatible with existing equipment and properly installed according to manufacturer guidelines.

Update Software and Firmware: If alarm system failures are attributed to software or firmware issues, update control system software, firmware, or programming to the latest versions provided by the manufacturer. Perform system diagnostics, error checks, and software validation tests to verify proper operation and functionality after software updates.

Test and Validate Repairs: Conduct comprehensive testing and validation procedures to verify that alarm system repairs have been successful and that all alarm functions are operating as intended. Simulate alarm activation scenarios, perform functional tests, and monitor system responses to ensure that alarms are triggered, transmitted, and received correctly.

Document Repairs and Maintenance: Maintain detailed records of alarm system repairs, component replacements, calibration procedures, and testing results for future reference and audit purposes. Document any changes made to alarm system configurations, settings, or software versions to facilitate troubleshooting and maintenance activities.

Implement Preventive Maintenance: Establish a preventive maintenance program for the lift station alarm system, including regular inspections, cleaning, calibration, and testing of alarm components. Schedule periodic maintenance tasks, such as sensor checks, battery replacements, and software updates, to prevent future alarm failures and ensure ongoing system reliability.

Training and Operator Awareness: Provide training to lift station operators and maintenance personnel on proper alarm system operation, troubleshooting techniques, and emergency response procedures. Educate operators about the importance of timely alarm detection, notification, and response to prevent equipment damage, system failures, and environmental incidents.

Boilers

PRESSURE PROBLEMS

Common pressure problems with boilers can arise due to various factors, affecting their efficiency, safety, and performance. Here are some typical pressure-related issues encountered with boilers:

Low Boiler Pressure: Low boiler pressure is a common problem that can result from several causes, including:

Water Leaks: Leaks in the boiler system, piping, valves, or fittings can lead to a loss of water and subsequent pressure drop.

Bleeding of Radiators: Air trapped in the radiators can prevent the proper circulation of water, resulting in reduced pressure in the boiler system.

Automatic Filling Valve Issues: Malfunctioning automatic filling valves may fail to maintain the required water level in the boiler, leading to low pressure.

Loss of Water Supply: Interruptions or restrictions in the water supply line can prevent adequate water replenishment in the boiler, causing pressure to drop.

Symptoms of low boiler pressure include weak heat output, cold radiators, and potential lockout or shutdown of the boiler's safety mechanisms.

High Boiler Pressure: High boiler pressure can occur due to various reasons, such as:

Overfilling: Excessive water intake caused by faulty automatic filling valves or manual overfilling can lead to elevated boiler pressure.

System Blockages: Blockages in the boiler system, piping, or heat exchanger can restrict water flow, causing pressure to build up.

Faulty Pressure Relief Valve: A malfunctioning pressure relief valve may fail to release excess pressure from the boiler, resulting in high pressure conditions.

Expansion Vessel Issues: An improperly sized or faulty expansion vessel can fail to absorb the expansion of water during heating, leading to pressure spikes. High boiler pressure can trigger safety concerns, including leaks, component damage, and potential rupture of the boiler tank.

Fluctuating Pressure: Fluctuating boiler pressure can be indicative of underlying issues, such as:

Waterlogged Expansion Vessel: If the expansion vessel is waterlogged or improperly charged, it may fail to accommodate changes in system volume, leading to pressure fluctuations.

Air Pockets: Air trapped within the boiler system can cause pressure variations as it accumulates or dissipates during operation.

Faulty Pressure Sensor: Malfunctioning pressure sensors or gauges may provide inaccurate readings, leading to perceived pressure fluctuations.

> **Fluctuating pressure can affect boiler efficiency, temperature control, and system stability, requiring prompt diagnosis and corrective action.**

Loss of Pressure Over Time: Gradual loss of pressure in the boiler system can occur due to slow leaks, evaporation, or bleeding of air from radiators over time. Corrosion, deteriorated seals, or loose connections in the boiler system may contribute to persistent pressure loss. Regular monitoring and periodic topping up of water levels may be necessary to maintain optimal pressure levels in the boiler.

> **Repairing pressure problems in boilers requires a systematic approach to diagnose the underlying causes and implement appropriate corrective measures. Here's a step-by-step guide on how to repair common pressure problems in boilers:**

Identify the Pressure Problem: Begin by determining whether the boiler is experiencing low pressure, high pressure, fluctuating pressure, or a gradual loss of pressure over time. Use pressure gauges, temperature readings, and visual inspection to assess the current pressure condition of the boiler system.

Check for Water Leaks: Inspect the boiler, piping, valves, fittings, and radiators for signs of water leaks, such as puddles, dampness, corrosion, or water stains. Repair any identified leaks by tightening connections, replacing faulty seals, or repairing damaged components to prevent further water loss and pressure drop.

Bleed Radiators (If Necessary): If low pressure is attributed to air trapped in the radiators, bleed the radiators to release trapped air and restore proper water circulation. Use a radiator key or bleed valve to open the air bleed screw at the top of each radiator until water begins to flow smoothly without air bubbles.

Inspect Automatic Filling Valve: Check the automatic filling valve (also known as the pressure reducing valve) for proper operation and adjustment. Ensure that the automatic filling valve is set to the correct pressure level and that it is functioning correctly to maintain the desired water level in the boiler system.

Service Pressure Relief Valve: Test the pressure relief valve to verify its functionality in releasing excess pressure from the boiler system. If the pressure relief valve is faulty or leaking, replace it with a new valve that meets the manufacturer's specifications and pressure rating.

Check Expansion Vessel: Inspect the expansion vessel for signs of damage, corrosion, or waterlogging. Recharge or replace the expansion vessel if necessary to ensure proper absorption of water expansion during heating cycles.

Address System Blockages: Clear any obstructions or blockages in the boiler system, piping, heat exchangers, or venting to restore proper water flow and pressure regulation. Use flushing or descaling procedures to remove limescale, mineral deposits, or sediment buildup that may be causing flow restrictions.

Adjust System Pressure: Use the boiler's pressure gauge or control panel to adjust the system pressure to the recommended operating range specified by the manufacturer. Avoid overfilling or over pressurizing the boiler system, as this can lead to safety hazards and damage to boiler components.

Test and Monitor: After making repairs and adjustments, test the boiler system to ensure that pressure levels are within the desired range and that all components are functioning correctly. Monitor the boiler system periodically to detect any recurring pressure problems and address them promptly to prevent further issues.

Regular Maintenance: Implement a regular maintenance schedule for the boiler system, including inspection, cleaning, and preventive maintenance tasks to prevent pressure problems and ensure long-term reliability.

WATER LEAKS

Water leaks in boilers can lead to various problems, including decreased system efficiency, property damage, and safety hazards. Identifying and addressing water leak issues promptly is crucial to prevent further damage and ensure the proper functioning of the boiler system. Here are some common water leak issues encountered with boilers:

Corroded or Damaged Components: Corrosion or deterioration of boiler components, such as heat exchangers, pipes, fittings, and seals, can result in water leaks. Corrosion may occur due to exposure to water, oxygen, chemicals, or high temperatures over time, leading to weakened or perforated surfaces and eventual leakage.

Faulty Seals and Gaskets: Seals and gaskets used in boiler components, including pump seals, valve seals, and access covers, may degrade or fail over time, resulting in water leaks. Installation, poor maintenance practices, or exposure to high temperatures can accelerate seal deterioration and contribute to leakage.

Pressure Relief Valve Leakage: Pressure relief valves are designed to release excess pressure from the boiler system to prevent overpressurization. If the pressure relief valve is faulty, improperly adjusted, or damaged, it may leak water continuously or intermittently, indicating a potential safety issue or system malfunction.

Expansion Tank Issues: Expansion tanks are used in boiler systems to accommodate fluctuations in water volume due to temperature changes during heating cycles. If the expansion tank is defective, improperly sized, or waterlogged, it may fail to absorb water expansion properly, leading to pressure fluctuations and potential leakage.

Condensate System Problems: Condensing boilers produce condensate as a byproduct of combustion, which is typically drained away through a condensate pipe or system. Clogs, blockages, or leaks in the condensate drain line, trap, or discharge point can cause condensate backup and leakage, potentially damaging boiler components and surrounding areas.

Pipe and Connection Leaks: Leaks can occur in boiler system piping, connections, joints, and fittings due to factors such as corrosion, mechanical stress, vibration, or improper installation. Water leaks from pipes or connections may manifest as drips, puddles, or wet spots in the vicinity of the boiler or along the piping route.

Faulty Heat Exchangers: Cracks, fractures, or deterioration in boiler heat exchangers can allow water to escape into the combustion chamber, burner compartment, or surrounding areas. Heat exchanger leaks may result from thermal expansion and contraction, metal fatigue, or chemical corrosion, necessitating repair or replacement of the affected components.

Condensation Leaks: Non-condensing boilers may experience condensation leaks if exhaust gases are not properly vented or if the flue gas temperature drops below the dew point. Condensation leaks may occur at the base of the boiler or near exhaust vents, indicating improper venting, inadequate insulation, or flue gas condensation issues.

> **Repairing water leak issues with boilers involves a systematic approach to identify the source of the leak and implement appropriate corrective measures. Here's a step-by-step guide on how to repair common water leak issues with boilers:**

Locate the Leak: Begin by identifying the source and location of the water leak in the boiler system. Inspect the boiler, piping, fittings, connections, seals, and components for signs of water leakage, such as drips, puddles, or wet spots.

Shut Down the Boiler: Prior to performing any repairs, shut down the boiler and turn off the power supply to ensure safety. Allow the boiler to cool down, if necessary, especially if the leak is associated with hot water or steam.

Inspect and Clean the Area: Clear away any debris, dirt, or obstructions around the area of the leak to facilitate inspection and repair. Use a flashlight or inspection mirror to examine concealed or hard-to-reach areas.

Repair or Replace Damaged Components: If the leak is originating from a specific component, such as a valve, fitting, seal, or gasket, assess the extent of the damage and determine whether repair or replacement is necessary. Repair minor leaks by tightening loose connections, replacing damaged seals or gaskets, or applying sealant or thread tape to threaded fittings. Replace severely damaged or corroded components with new parts that meet the manufacturer's specifications and compatibility requirements.

Service Pressure Relief Valve: If the pressure relief valve is leaking, inspect it for signs of damage, corrosion, or debris buildup. Clean the valve and check for proper operation. Adjust the pressure relief valve settings according to manufacturer recommendations, if necessary, to prevent overpressurization and leakage.

Address Expansion Tank Issues: Inspect the expansion tank for signs of waterlogging, corrosion, or damage. Drain the expansion tank and check the air pressure using a tire pressure gauge. Recharge or replace the expansion tank bladder as needed to restore proper pressure and function.

Clear Condensate Drain Lines: If the leak is related to condensate drainage issues, clear any clogs, blockages, or debris from the condensate drain line, trap, or discharge point. Use a drain snake, brush, or compressed air to remove obstructions and ensure proper condensate flow.

Check Heat Exchanger and Flue: Inspect the boiler heat exchanger and flue for signs of cracks, corrosion, or damage that may be causing leaks. Repair or replace damaged heat exchanger sections or flue components to prevent further leakage and ensure safe operation.

Test and Monitor: After completing repairs, test the boiler system to verify that the leak has been resolved and that all components are functioning correctly. Monitor the boiler system periodically for any signs of recurrence or new leaks and address them promptly to prevent further issues.

Schedule Regular Maintenance: Implement a regular maintenance schedule for the boiler system, including inspection, cleaning, and preventive maintenance tasks, to prevent water leak issues and ensure long-term reliability.

LIMESCALE AND MINERAL DEPOSITS

Limescale and mineral deposits are common issues in boiler systems, particularly in areas with hard water. These deposits can accumulate on internal surfaces, such as heat exchangers, pipes, and valves, leading to reduced efficiency, increased energy consumption, and potential damage to boiler components. Here's how limescale and mineral deposits can affect boilers and how to address the issues:

EFFECTS OF LIMESCALE AND MINERAL DEPOSITS:

Reduced Heat Transfer: Limescale and mineral deposits act as insulators, inhibiting heat transfer from the boiler's heating elements to the water. This reduces the boiler's efficiency and increases energy consumption.

Increased Operating Costs: As the boiler works harder to maintain desired temperatures, energy costs can rise significantly due to the buildup of limescale and mineral deposits.

Decreased Lifespan: Accumulation of deposits can lead to overheating of components, increased corrosion, and potential damage to boiler surfaces, ultimately reducing the lifespan of the boiler.

System Blockages: Heavy buildup of limescale and mineral deposits can cause blockages in pipes, valves, and heat exchangers, resulting in reduced water flow, pressure issues, and potential system failures.

ADDRESS LIMESCALE AND MINERAL DEPOSITS:

Water Softening: Install a water softening system to treat hard water before it enters the boiler. Water softeners remove calcium and magnesium ions, which are the primary components of limescale.

Chemical Descaling: Periodically descale the boiler using chemical descaling agents specifically designed for boilers. These agents dissolve limescale and mineral deposits, allowing them to be flushed out of the system.

Mechanical Cleaning: Use mechanical methods such as brushing, scraping, or power flushing to remove limescale and mineral deposits from boiler components. This may require disassembly of certain parts for thorough cleaning.

Magnetic Water Treatment: Install magnetic water treatment devices on the boiler's water supply line. These devices use magnets to alter the physical properties of minerals in the water, reducing their ability to form limescale.

Scale Inhibitors: Add scale inhibitors or corrosion inhibitors to the boiler's water supply to prevent the formation of limescale and mineral deposits. These chemicals can help minimize buildup and extend the life of the boiler.

Regular Maintenance: Implement a regular maintenance schedule for the boiler, including inspection, cleaning, and descaling procedures. Regular maintenance can help prevent limescale and mineral deposits from accumulating and ensure optimal boiler performance.

Repairing limescale and mineral deposits in a boiler involves several steps to remove existing buildup and prevent future accumulation. Here's a detailed guide on how to repair limescale and mineral deposits in a boiler:

Assess the Extent of Deposits: Begin by assessing the extent of limescale and mineral deposits in the boiler system. Inspect boiler components such as heat exchangers, pipes, valves, and tanks for visible signs of scaling, including white or yellowish buildup.

Shut Down the Boiler: Prior to performing any repairs, shut down the boiler and turn off the power supply to ensure safety. Allow the boiler to cool down if necessary, especially if the descaling process involves hot water or steam.

Drain the Boiler: Drain the boiler system to remove any remaining water and sediment. Follow manufacturer guidelines or consult with a professional to ensure proper draining procedures are followed.

Chemical Descale: Prepare a descaling solution according to the manufacturer's instructions or use a commercially available descaling agent specifically designed for boilers. Introduce the descaling solution into the boiler system through the appropriate access points, such as the fill valve or relief valve connection. Allow the descaling solution to circulate through the boiler system for the recommended duration to dissolve limescale and mineral deposits. Flush the boiler system thoroughly with clean water to remove dissolved deposits and residual descaling solution.

Mechanical Cleaning: For stubborn deposits or hard-to-reach areas, use mechanical methods such as brushing, scraping, or power flushing to dislodge and remove limescale and mineral buildup. Use boiler-specific brushes, scrapers, or power flushing equipment to clean heat exchanger tubes, pipe interiors, and other components affected by deposits.

Inspect and Replace Damaged Components: Inspect boiler components for signs of damage, corrosion, or deterioration caused by limescale and mineral deposits. Replace any damaged seals, gaskets, or parts that may have been compromised due to scaling or corrosion to ensure proper sealing and functionality.

Install Prevention Measures: Implement preventive measures to minimize the recurrence of limescale and mineral deposits in the boiler system. Consider installing water softening systems, magnetic water treatment devices, scale inhibitors, or corrosion inhibitors to prevent scale formation and protect boiler components.

Regular Maintenance: Establish a regular maintenance schedule for the boiler system, including periodic descaling procedures, inspection, and cleaning. Monitor water quality, flow rates, and boiler performance regularly to detect and address any signs of scaling or deposit buildup promptly.

Follow Manufacturer Guidelines: Follow manufacturer recommendations, guidelines, and safety precautions when performing boiler repairs, descaling procedures, and maintenance activities.

IGNITION AND COMBUSTION ISSUES

Common ignition and combustion issues in a boiler can result in inefficient operation, increased fuel consumption, and potentially hazardous conditions. Here are some typical problems associated with ignition and combustion in boilers:

Failure to Ignite: A boiler failing to ignite is a common issue that can occur due to various reasons, including:

Faulty Ignition Components: Problems with ignition electrodes, spark plugs, ignition transformers, or ignition control modules can prevent the boiler from igniting.

Fuel Supply Issues: Insufficient gas pressure, air or gas flow restrictions, or fuel valve malfunctions can hinder proper fuel delivery and ignition.

Combustion Air Supply Problems: Inadequate combustion air intake or blocked air vents can prevent proper air-fuel mixture and ignition.

Failure to ignite may result in repeated ignition attempts, delayed heating, or boiler lockout, leading to discomfort and inconvenience.

Intermittent Ignition: Intermittent ignition refers to situations where the boiler ignites inconsistently or sporadically during operation. Causes of intermittent ignition may include intermittent gas supply, poor electrode positioning, dirty ignition components, or improper combustion air adjustment. Intermittent ignition can lead to unreliable heating, temperature fluctuations, and increased wear on ignition components.

Delayed Ignition: Delayed ignition occurs when the ignition of air-fuel mixture is delayed after the burner is activated, resulting in a noticeable delay in heating. Common causes of delayed ignition include dirty burners, clogged pilot orifices, excessive gas flow rates, or improper burner adjustment. Delayed ignition can produce loud booming noises, increased fuel consumption, and potential damage to boiler components due to repeated ignition attempts.

Uneven Flame Patterns: Uneven flame patterns in the boiler burner indicate irregular combustion and can result from factors such as:

Dirty or Misaligned Burner Components: Accumulation of dirt, debris, or soot on burner ports, diffusers, or baffles can disrupt airflow and fuel distribution, leading to uneven flames.

Improper Air-Fuel Ratio: Incorrect adjustment of air dampers, fuel valves, or combustion control settings can cause imbalances in the air-fuel mixture and uneven combustion. Uneven flame patterns can reduce boiler efficiency, increase emissions, and contribute to heat exchanger fouling.

Sooting and Carbon Buildup: Sooting and carbon buildup on boiler surfaces, burner components, and heat exchangers can occur due to incomplete combustion or improper burner operation. Factors contributing to sooting include inadequate combustion air supply, improper air-fuel mixture, low combustion chamber temperatures, or high excess air levels. Sooting can reduce heat transfer efficiency, impair burner performance, and pose safety risks, including carbon monoxide (CO) emissions.

Flame Rollout: Flame rollout occurs when flames from the burner extend beyond the combustion chamber or heat exchanger, posing a risk of ignition of nearby combustible materials. Causes of flame rollout may include blocked flue passages, improper venting, inadequate combustion air supply, or damaged heat exchanger seals. Flame rollout can result in equipment damage, safety hazards, and potential fire hazards, requiring immediate attention and corrective action.

Repairing ignition and combustion issues in a boiler requires careful troubleshooting and corrective action to address underlying problems effectively. Here's a step-by-step guide on how to repair common ignition and combustion issues in a boiler:

Safety Precautions: Before performing any repairs, ensure the boiler is shut down properly and the power supply is turned off. Follow all safety procedures outlined in the boiler's manual and wear appropriate personal protective equipment (PPE).

Inspect Ignition Components: Begin by inspecting ignition components such as electrodes, spark plugs, ignition transformers, and control modules for signs of damage, wear, or corrosion. Clean the ignition electrodes and ensure proper positioning to facilitate consistent spark ignition. Test the continuity of ignition transformers and control modules using a multimeter to verify functionality.

Check Fuel Supply: Verify that the boiler has an adequate supply of fuel (e.g., natural gas, propane, oil) and that fuel valves are open and functioning correctly. Test gas pressure using a manometer to ensure it meets the manufacturer's specifications. Adjust gas pressure regulators if necessary.

Inspect Combustion Air Supply: Check the combustion air intake for obstructions, blockages, or restrictions that may be impeding airflow to the burner. Clean air intake vents, louvers, and screens to ensure proper combustion air supply.

Clean Burner Components: Remove burner assemblies and inspect for signs of sooting, carbon buildup, or debris accumulation on burner ports, diffusers, and baffles. Use a wire brush, compressed air, or a vacuum cleaner to clean burner components thoroughly and restore proper airflow and fuel distribution.

Adjust Air-Fuel Ratio: Check the air-fuel ratio settings and adjust combustion air dampers, fuel valves, or combustion control parameters as needed to achieve optimal combustion. Monitor flame patterns and combustion characteristics to ensure a stable, blue flame with minimal sooting or flame rollout.

Test Ignition Sequence: Manually initiate the ignition sequence and observe the ignition process to ensure consistent sparking, ignition, and flame establishment. Verify that ignition electrodes are generating sparks, ignition transformers are delivering sufficient voltage, and control modules are initiating ignition sequences correctly.

Inspect Flue and Venting: Inspect flue passages, venting systems, and draft regulators for blockages, restrictions, or damage that may affect proper exhaust gas ventilation. Clean flue passages and ensure proper venting to prevent backdrafting, flame rollout, or carbon monoxide buildup.

Monitor Combustion Performance: Monitor combustion performance using combustion analyzers, gas analyzers, or flame sensors to assess combustion efficiency, CO levels, and other combustion parameters. Make adjustments to burner operation, combustion controls, or system settings to optimize combustion performance and minimize emissions.

Perform System Test: After making repairs and adjustments, restart the boiler and perform a comprehensive system test to verify proper ignition, combustion, and heating operation. Monitor boiler performance, flame stability, temperature control, and safety interlocks to ensure reliable operation.

Schedule Regular Maintenance: Establish a regular maintenance schedule for the boiler, including inspection, cleaning, and preventive maintenance tasks to prevent ignition and combustion issues. Keep detailed records of maintenance activities, repairs, and performance assessments for future reference and troubleshooting.

THERMOSTAT MALFUNCTION

Thermostat malfunctions in boilers can disrupt heating operation, lead to temperature fluctuations, and impact energy efficiency. Here are some common thermostat malfunction issues encountered with boilers:

Inaccurate Temperature Readings: One of the most common thermostat issues is inaccurate temperature readings, where the thermostat displays temperatures that do not reflect the actual room temperature. This can occur due to sensor calibration errors, aging components, or placement of the thermostat in an inappropriate location (e.g., near heat sources or drafts).

Failure to Respond to Settings: Thermostats may fail to respond to temperature settings or commands, preventing the boiler from activating or deactivating as desired. This issue can stem from electrical faults, mechanical failures, or programming errors within the thermostat.

Short Cycling: Short cycling refers to frequent and rapid cycling of the boiler, where it turns on and off more frequently than necessary. Thermostat malfunctions, such as faulty temperature sensors or wiring issues, can contribute to short cycling, leading to energy wastage, increased wear on components, and reduced efficiency.

Constantly Running Boiler: A malfunctioning thermostat may cause the boiler to run continuously, even when the desired temperature has been reached. This can result from thermostat wiring faults, malfunctioning relays, or improper thermostat settings.

No Response to User Input: Sometimes, thermostats may fail to respond to user input, such as changes in temperature settings or mode adjustments. This issue can arise due to software glitches, battery issues (in battery-operated thermostats), or communication errors between the thermostat and the boiler control unit.

Intermittent Operation: Intermittent operation occurs when the thermostat intermittently loses communication with the boiler or experiences intermittent electrical faults. This can lead to unpredictable heating cycles, inconsistent temperature control, and discomfort for occupants.

Dead or Nonresponsive Thermostat: In some cases, thermostats may become completely nonresponsive, failing to power on or display any information. This can be due to power supply issues, battery depletion (in battery-operated thermostats), or internal component failures.

Display Errors: Display errors on the thermostat, such as flickering screens, error codes, or blank displays, can indicate underlying electrical issues, software glitches, or component failures.

Incorrect Programming: Incorrect thermostat programming, such as incorrect temperature setpoints, schedules, or operating modes, can lead to suboptimal heating performance and energy wastage.

Addressing thermostat malfunction issues in boilers requires thorough troubleshooting and corrective action. Here's how to repair common thermostat malfunctions:

Check Power Supply: Ensure that the thermostat is receiving power from the mains or batteries. Replace batteries if necessary and check circuit breakers or fuses if the thermostat is hardwired.

Inspect Wiring Connections: Inspect the wiring connections between the thermostat and the boiler control unit. Tighten loose connections and replace damaged or corroded wires.

Calibrate Temperature Sensors: If the thermostat displays inaccurate temperature readings, recalibrate or replace temperature sensors as needed to ensure accurate temperature measurement.

Reset or Reboot Thermostat: Reset or reboot the thermostat according to the manufacturer's instructions to clear any temporary glitches or software errors.

Update Firmware (If Applicable): If the thermostat is programmable or connected to a smart home system, ensure that the firmware is up to date to resolve any known software issues. Adjust Settings and Programming: Verify and adjust thermostat settings, temperature setpoints, and operating modes to meet desired comfort levels and optimize energy efficiency.

Replace Thermostat: If troubleshooting fails to resolve the issue, consider replacing the thermostat with a new unit compatible with the boiler system.

Repairing thermostat malfunctions in boilers typically involves a combination of troubleshooting steps to identify the underlying issue and implementing appropriate corrective measures. Here's a step-by-step guide on how to repair common thermostat malfunction issues in boilers:

Verify Power Supply: Check the power supply to the thermostat to ensure it is receiving power. If the thermostat is battery-operated, replace the batteries with fresh ones and ensure they are installed correctly. For hardwired thermostats, check the circuit breaker or fuse to ensure it hasn't tripped.

Inspect Wiring Connections: Turn off the power to the boiler and thermostat before inspecting wiring connections. Ensure that all wires are securely connected to their respective terminals on both the thermostat and the boiler control unit. Tighten any loose connections and replace any damaged wires.

Calibrate Temperature Sensors: If the thermostat is displaying inaccurate temperature readings, recalibrate or replace the temperature sensors. Follow the manufacturer's instructions for recalibration or replacement procedures.

Reset or Reboot Thermostat: Many thermostats have a reset function that can help clear temporary glitches or errors. Refer to the thermostat's manual for instructions on how to perform a reset. Alternatively, you can try rebooting the thermostat by removing the batteries or power supply for a few minutes before reconnecting them.

Check for Firmware Updates: If the thermostat is programmable or connected to a smart home system, check if there are any available firmware updates. Updating the firmware may resolve software-related issues. Follow the manufacturer's instructions for updating the firmware.

Adjust Settings and Programming: Verify that the thermostat settings, temperature setpoints, and programming schedules are correctly configured. Make any necessary adjustments to ensure they meet your comfort preferences and heating requirements.

Test Thermostat Operation: Turn on the power to the boiler and observe the thermostat's operation. Verify that it responds appropriately to temperature changes and commands to activate or deactivate the boiler as expected.

Replace Thermostat: If troubleshooting steps fail to resolve the issue, consider replacing the thermostat with a new one. Choose a thermostat that is compatible with your boiler system and meets your heating control needs.

Consult with Professionals: If you are unable to diagnose or repair the thermostat malfunction on your own, or if the boiler system requires specialized knowledge or tools, it's best to consult with qualified HVAC technicians or boiler service professionals for assistance.

Perform Regular Maintenance: To prevent future thermostat malfunctions and ensure optimal performance of your boiler system, perform regular maintenance, including cleaning, calibration, and inspection of thermostat components.

FLUE GAS AND VENTILATION

Flue gas and ventilation issues in boilers can lead to safety hazards, inefficient operation, and increased emissions of harmful gases. Here are some common problems associated with flue gas and ventilation in boilers:

Incomplete Combustion: Incomplete combustion occurs when the boiler does not burn fuel completely, resulting in the production of carbon monoxide (CO) and other harmful gases. Causes include insufficient oxygen supply, improper air-fuel ratio, dirty burners, or flue gas recirculation. Incomplete combustion can pose serious health risks and is a sign of inefficient operation.

Backdrafting: Backdrafting occurs when flue gases flow back into the boiler or living space instead of being expelled through the flue vent. Causes include negative pressure within the building, inadequate chimney draft, blockages in the flue or vent, or improper venting design. Backdrafting can lead to CO buildup and exposure risks.

Flue Gas Leakage: Flue gas leakage can occur due to cracks, gaps, or corrosion in the boiler's heat exchanger or flue venting system. Leakage allows combustion gases, including CO, to escape into the boiler room or living space, posing health and safety hazards. It can also reduce boiler efficiency and increase fuel consumption.

Flue Gas Condensation: Flue gas condensation occurs when water vapor in flue gases condenses into liquid form due to low flue gas temperatures. Causes include high-efficiency boiler operation, cold flue venting, or improper vent sizing. Condensation can lead to corrosion of boiler components and flue systems, as well as chimney deterioration.

Blockages and Restrictions: Blockages or restrictions in the flue venting system can impede the flow of combustion gases, leading to pressure buildup, reduced boiler efficiency, and potential safety hazards. Common sources of blockages include debris, bird nests, ice buildup, or structural damage to the chimney or vent.

Improper Ventilation Design: Inadequate ventilation design can result in insufficient air supply for combustion and ventilation of flue gases. Poor ventilation design can lead to restricted airflow, pressure imbalances, and backdrafting. It may also contribute to CO buildup and indoor air quality issues.

Chimney Height and Location: Chimney height and location play a crucial role in ensuring proper draft and effective expulsion of flue gases. Chimneys located too close to neighboring structures or obstacles may experience downdrafts or poor draft conditions, leading to backdrafting or inadequate venting.

Exhaust Gas Temperature: Exhaust gas temperature should be within the optimal range to prevent flue gas condensation and ensure efficient boiler operation. High exhaust gas temperatures can lead to energy wastage, while low temperatures may cause condensation and corrosion issues.

Flue Gas Analysis: Regular flue gas analysis is essential for monitoring combustion efficiency, emissions, and safety. Analyzing flue gases helps identify combustion problems, adjust air-fuel ratios, and ensure compliance with environmental regulations.

Addressing flue gas and ventilation issues in boilers requires careful diagnosis and corrective action to ensure safe and efficient operation. Here's how to repair common problems:

Inspect Flue Venting System: Inspect the flue venting system for blockages, restrictions, or damage. Clear any debris, nests, or ice buildup from the chimney or vent pipes. Repair or replace damaged sections as needed.

Check for Flue Gas Leakage: Perform a visual inspection of the boiler's heat exchanger and flue connections for signs of corrosion, cracks, or gaps. Seal any leaks with high-temperature silicone or other appropriate sealants.

Optimize Combustion: Adjust the air-fuel ratio and combustion settings to achieve complete combustion and minimize CO emissions. Clean burners, adjust combustion air dampers, and ensure proper ventilation to optimize combustion efficiency.

Improve Ventilation: Improve ventilation in the boiler room or installation area to ensure an adequate air supply for combustion and ventilation of flue gases. Install additional air vents or louvers if necessary to reduce negative pressure.

Address Condensation Issues: Increase flue gas temperatures by adjusting boiler settings, installing a flue gas economizer, or insulating flue venting to prevent condensation. Ensure proper vent sizing and slope to facilitate drainage of condensate.

Verify Chimney Height and Location: Evaluate the chimney height, location, and surrounding structures to ensure proper draft conditions and prevent downdrafts or backdrafting. Consider extending the chimney height or relocating the chimney if necessary.

Perform Flue Gas Analysis: Conduct regular flue gas analysis to monitor combustion efficiency, CO levels,

PUMP FAILURE

Pump failure in boilers can disrupt the heating system, leading to loss of heat, potential damage to the boiler, and discomfort for occupants. Here are some common pump failure issues in boilers:

Mechanical Wear and Tear: Over time, the pump's mechanical components, such as bearings, seals, impellers, and motor shafts, can wear out due to normal usage. Wear and tear can result in decreased pump performance, reduced efficiency, and eventual pump failure.

Blockages and Clogs: Pump impellers and inlet/outlet ports can become clogged with debris, sludge, or mineral deposits over time. Blockages restrict water flow through the pump, causing increased pressure, reduced circulation, and potential pump overheating and failure.

Air Locks: Air locks occur when air becomes trapped in the pump or piping system, preventing the pump from priming or circulating water effectively. Air locks can impede pump operation, decrease flow rates, and lead to pump cavitation and damage.

Electrical Issues: Electrical problems, such as motor failures, wiring faults, or control circuit malfunctions, can cause pump failure. Electrical issues may result from overheating, moisture ingress, power surges, or component degradation.

Lubrication Problems: Inadequate lubrication of pump bearings and moving parts can lead to increased friction, overheating, and premature wear. Lack of proper lubrication maintenance can cause bearing failure and result in pump malfunction.

Seal Leaks: Pump seals can deteriorate over time, leading to leaks and loss of water pressure. Seal leaks can result from wear, corrosion, or improper installation, causing water damage to pump components and surrounding areas.

Overheating: Overheating of the pump motor can occur due to prolonged operation under high load conditions, inadequate cooling, or insufficient lubrication. Overheating can lead to motor failure, winding insulation damage, or thermal shutdown, causing the pump to stop working.

Incorrect Sizing: Using an undersized or oversized pump for the boiler system can lead to inefficiency, excessive wear, and premature failure. Incorrectly sized pumps may struggle to meet system demands or operate outside their optimal performance range, leading to reliability issues.

Corrosion and Erosion: Corrosion and erosion of pump components, such as impellers, casing, and piping, can occur due to exposure to corrosive water, chemicals, or high-velocity flow. Corrosion and erosion weaken pump parts, decrease efficiency, and increase the risk of mechanical failure.

Vibration and Misalignment: Excessive vibration or misalignment of pump components, shafts, or motor couplings can lead to premature wear, seal leaks, and bearing failure. Vibration and misalignment issues may result from improper installation, poor maintenance, or structural problems.

Addressing pump failure issues in boilers requires thorough diagnosis and appropriate corrective action. Here's how to repair common pump failure issues:

Inspect and Clean: Inspect the pump and surrounding piping for blockages, debris, or air locks. Clean the pump impeller, inlet, and outlet ports to remove any obstructions.

Check Electrical Connections: Inspect electrical connections, wiring, and control circuits for signs of damage, corrosion, or loose connections. Repair or replace faulty components as needed.

Prime the Pump: If the pump is not priming or circulating water properly, bleed air from the system to eliminate air locks. Ensure proper priming and venting to allow the pump to operate efficiently.

Replace Seals and Bearings: Replace worn or damaged pump seals and bearings to prevent leaks and restore proper pump operation. Use OEM replacement parts and follow manufacturer recommendations.

Lubricate Moving Parts: Lubricate pump bearings and moving parts with the appropriate lubricants to reduce friction, prevent overheating, and extend component life.

Address Overheating: Improve pump cooling by ensuring adequate airflow and ventilation around the pump motor. Address any factors contributing to overheating, such as high ambient temperatures or restricted airflow.

Correct Sizing: Ensure the pump is correctly sized for the boiler system's flow rate, pressure requirements, and operating conditions. Replace undersized or oversized pumps with appropriately sized ones.

Protect Against Corrosion: Implement corrosion prevention measures, such as using corrosion-resistant materials, applying protective coatings, or installing sacrificial anodes to protect pump components from corrosion and erosion.

Balance and Align: Balance pump impellers and ensure proper alignment of pump shafts, bearings, and motor couplings to minimize vibration, reduce wear, and prolong component life.

Monitor and Maintain: Regularly monitor pump performance, temperature, and vibration levels. Implement a preventive maintenance program to inspect, lubricate, and service the pump at scheduled intervals.

ELECTRICAL AND CONTROL SYSTEM PROBLEMS

Electrical and control system problems in boilers can disrupt operation, compromise safety, and lead to equipment damage. Here are some common issues encountered with electrical and control systems in boilers:

Faulty Wiring: Faulty or damaged wiring can result in intermittent power supply, short circuits, or electrical arcing. Wiring problems may occur due to wear, corrosion, rodents, or poor installation practices.

Control Panel Malfunctions: Control panel malfunctions, such as malfunctioning relays, switches, or indicators, can prevent proper control and operation of the boiler system. Issues may arise from electrical failures, component degradation, or moisture ingress.

Sensor and Transmitter Failures: Sensors and transmitters, including temperature sensors, pressure sensors, and flow meters, are critical for monitoring boiler parameters and controlling system operation. Failures in these components can lead to inaccurate readings, improper control, and system malfunctions.

Power Supply Problems: Power supply issues, such as voltage fluctuations, power surges, or phase imbalance, can affect the operation of electrical components in the boiler system. Inadequate power supply can cause erratic behavior, equipment damage, or complete shutdown.

Burner and Ignition System Failures: Problems with burner and ignition systems, including faulty electrodes, ignition transformers, or flame sensors, can prevent proper combustion and ignition of fuel. Electrical faults or component degradation may contribute to burner malfunctions.

Boiler Control System Errors: Errors or malfunctions in the boiler control system, such as programming errors, software glitches, or communication failures, can disrupt system operation and control. Control system issues may arise from hardware or software problems.

Safety Interlock Failures: Safety interlocks, such as pressure switches, temperature limits, and flame safeguards, are designed to prevent unsafe operating conditions and protect against hazards. Failure of safety interlocks can compromise boiler safety and lead to potential accidents.

Electrical Overload: Electrical overload occurs when the electrical system draws more current than it is designed to handle, leading to overheating, equipment damage, or tripped circuit breakers. Overload may result from excessive loads, short circuits, or insulation breakdown.

Grounding and Bonding Issues: Improper grounding and bonding can result in electrical hazards, equipment damage, and interference with control signals. Inadequate grounding may lead to stray currents, electromagnetic interference, or shock hazards.

Communication Failures: Communication failures between boiler components, control devices, or external systems can disrupt data exchange and control functions. Issues with communication protocols, wiring, or network connectivity may affect system integration and operation.

Transformer and Capacitor Problems: Transformers and capacitors used in electrical control circuits can degrade over time or fail due to electrical stress, overheating, or component aging. Malfunctioning transformers or capacitors may cause voltage irregularities or control system malfunctions.

> **Addressing electrical and control system problems in boilers requires systematic troubleshooting and corrective action. Here's how to repair common issues:**

Inspect and Test Wiring: Inspect wiring connections, terminals, and junction boxes for signs of damage, corrosion, or loose connections. Test continuity, insulation resistance, and voltage levels to identify wiring faults.

Check Control Panel Components: Inspect control panel components, such as relays, switches, indicators, and circuit boards, for visible damage or signs of overheating. Test functionality and replace faulty components as needed.

Calibrate Sensors and Transmitters: Calibrate temperature sensors, pressure transmitters, and other instrumentation to ensure accurate readings and proper control. Replace malfunctioning sensors or transmitters if calibration fails to resolve issues.

Stabilize Power Supply: Install voltage regulators, surge protectors, or power conditioning devices to stabilize the power supply and protect sensitive equipment from voltage fluctuations or power surges.

Maintain Burner and Ignition Systems: Clean and inspect burner components, ignition electrodes, transformers, and flame sensors regularly to ensure proper operation and combustion. Replace worn or damaged parts to prevent ignition failures.

Verify Control System Settings: Review and verify control system settings, programming parameters, and setpoints to ensure correct operation and control logic. Reset or reprogram the control system as needed to resolve errors or glitches.

Test Safety Interlocks: Test safety interlocks, pressure switches, temperature limits, and flame safeguards to verify proper operation and response to unsafe conditions. Repair or replace faulty interlocks to restore safety functionality.

Balance Electrical Loads: Balance electrical loads across phases, circuits, and components to prevent overload conditions and ensure uniform distribution of power. Install additional circuits or redistribute loads as necessary to alleviate overloading.

Improve Grounding and Bonding: Enhance grounding and bonding systems to reduce electrical hazards and improve system reliability. Ensure proper grounding connections, bonding conductors, and equipotential bonding to minimize stray currents and interference.

Diagnose Communication Issues: Diagnose communication failures using diagnostic tools, protocol analyzers, or network testers to identify root causes. Check wiring connections, network configurations, and device settings to resolve communication problems.

Replace Faulty Components: Replace faulty transformers, capacitors, relays, or control boards with compatible replacements from the manufacturer. Follow proper installation procedures and ensure compatibility with existing system components.

Perform Regular Maintenance: Implement a preventive maintenance program to inspect, clean, and test electrical and control system components at scheduled intervals. Address potential issues proactively to prevent downtime and ensure reliable operation.

FUEL SUPPLY AND COMBUSTION AIR ISSUES

Issues related to fuel supply and combustion air in boilers can significantly impact combustion efficiency, heating performance, and safety. Here are some common problems encountered with fuel supply and combustion air in boilers:

Insufficient Fuel Supply: Insufficient fuel supply can lead to incomplete combustion, reduced heat output, and inefficient boiler operation. Causes may include fuel line blockages, low fuel pressure, or fuel valve malfunctions.

Fuel Contamination: Contaminants such as dirt, water, or debris in the fuel supply can disrupt combustion, clog fuel nozzles, and damage burner components. Contamination may occur during fuel storage, handling, or transportation.

Incorrect Fuel Type: Using the wrong type of fuel or fuel with incorrect specifications can result in poor combustion, excessive soot buildup, and equipment damage. Ensure that the boiler is designed and configured to use the specified fuel type.

Air-Fuel Ratio Imbalance: Improper air-fuel ratio can lead to inefficient combustion, excess emissions, and reduced boiler efficiency. Insufficient combustion air or excessive fuel supply can result in incomplete combustion and carbon monoxide (CO) production.

Inadequate Combustion Air Supply: Inadequate combustion air supply restricts oxygen availability for combustion, leading to reduced combustion efficiency, increased fuel consumption, and elevated emissions. Factors such as insufficient ventilation or blocked air intakes can contribute to this issue.

Combustion Air Leakage: Combustion air leaks into or out of the boiler system can disrupt combustion air supply, affect burner performance, and compromise boiler efficiency. Leaks may occur at air duct connections, burner openings, or boiler seals.

Excessive Combustion Air: Excessive combustion air can lead to overdilution of fuel-air mixture, reduced flame temperature, and inefficient combustion. This may occur due to improper air damper settings, oversized combustion air fans, or poor air distribution.

Fluctuating Combustion Air Temperature: Fluctuations in combustion air temperature can affect combustion stability, flame characteristics, and boiler efficiency. Variations in ambient air temperature or air preheating systems can impact combustion air temperature.

Air Infiltration: Air infiltration into the boiler room from external sources can disrupt combustion air supply, affect burner operation, and compromise combustion efficiency. Seal leaks, openings, or gaps in the boiler room to prevent unwanted air infiltration.

High Flue Gas Temperatures: High flue gas temperatures can indicate inefficient combustion, heat loss, or excess air in the combustion process. Optimize combustion parameters, adjust air-fuel ratio, and improve insulation to reduce flue gas temperatures.

Draft Problems: Draft problems, such as inadequate chimney draft or negative pressure in the boiler room, can affect combustion air supply and venting of flue gases. Address chimney height, size, or draft inducer systems to ensure proper draft.

Erosion and Corrosion: Erosion and corrosion of burner components, combustion air ducts, and flue systems can occur due to exposure to corrosive gases, high temperatures, or abrasive particles. Regular inspection and maintenance are essential to prevent damage.

> **To address fuel supply and combustion air issues in boilers, consider the following repair and maintenance actions:**

Inspect and Clean Fuel System: Regularly inspect and clean fuel storage tanks, filters, strainers, and fuel lines to remove contaminants and ensure uninterrupted fuel supply.

Verify Fuel Specifications: Confirm that the boiler is supplied with the correct type of fuel, meeting specified quality standards and compatibility requirements.

Adjust Air-Fuel Ratio: Adjust air and fuel flow rates to achieve the optimal air-fuel ratio for combustion. Calibrate combustion control systems and burner settings accordingly.

Ensure Proper Ventilation: Provide adequate ventilation in the boiler room to ensure sufficient combustion air supply and proper air distribution to burners.

Seal Air Leaks: Seal air leaks in combustion air ducts, burner openings, and boiler enclosures to prevent air infiltration and maintain combustion efficiency.

Install Combustion Air Preheaters: Consider installing combustion air preheaters to improve combustion efficiency and reduce energy consumption by recovering waste heat from flue gases.

Optimize Combustion Parameters: Monitor and adjust combustion parameters such as excess air levels, stack temperatures, and oxygen concentrations to optimize combustion efficiency and minimize emissions.

Perform Regular Maintenance: Implement a comprehensive maintenance program to inspect, clean, and tune boiler combustion systems at scheduled intervals. Replace worn or damaged components as needed to maintain optimal performance.

Conduct Combustion Analysis: Periodically perform combustion analysis to assess combustion efficiency, flue gas composition, and emissions levels. Use the results to optimize combustion settings and troubleshoot performance issues.

Swimming Pools

Algae growth is a common issue in swimming pools, particularly in warmer climates or when pool maintenance practices are inadequate. Algae can proliferate rapidly, causing water discoloration, unpleasant odors, and potential health hazards. Here are some common algae growth issues encountered in swimming pools:

Green Algae: Green algae (phylum: Chlorophyta) are the most common type of algae found in swimming pools. They typically appear as green patches or streaks on pool surfaces and can quickly spread if not addressed promptly.

Black Algae: Black algae (kingdom: Plantae, phylum: Ochrophyta) are dark green to black in color and commonly form as small dots or spots on pool walls, steps, and surfaces. Black algae have strong root systems that make them difficult to eradicate.

Mustard (Yellow) Algae: Mustard algae (phylum: Ochrophyta) are yellowish green in color and often appear as slimy deposits on pool walls, steps, and shaded areas. They can be challenging to eliminate and may return if not treated effectively.

Pink Algae: Pink algae (bacteria: Serratia marcescens) are actually bacteria rather than true algae. They appear as pinkish or reddish-colored slime deposits on pool surfaces, especially in areas with low circulation and sunlight exposure.

Factors contributing to algae growth:

Insufficient Sanitizer Levels: Low chlorine or sanitizer levels allow algae to thrive in pool water.

Poor Water Circulation: Inadequate filtration and circulation can create stagnant areas where algae can grow.

Imbalanced Water Chemistry: Poorly balanced pH, alkalinity, or calcium hardness levels can promote algae growth.

Warm Water Temperatures: Warmer water temperatures accelerate algae growth, especially during the summer months.

Sunlight Exposure: Direct sunlight can stimulate algae growth by providing energy for photosynthesis.

Organic Contaminants: Organic debris, such as leaves, grass, or body oils, provide nutrients for algae growth.

Inadequate Pool Maintenance: Irregular cleaning, brushing, and vacuuming allow algae to accumulate and proliferate.

> **To address algae growth issues in swimming pools, consider the following preventive measures and corrective actions:**

Maintain Proper Sanitizer Levels: Ensure that chlorine or other sanitizers are maintained within the recommended range to prevent algae growth. Shock the pool periodically to boost sanitizer levels and eliminate organic contaminants.

Optimize Filtration and Circulation: Run the pool pump and filter for an adequate duration each day to achieve proper water turnover and circulation. Clean or backwash the filter regularly to remove algae spores and debris.

Balance Water Chemistry: Test pool water regularly and adjust pH, alkalinity, and calcium hardness levels as needed to maintain proper balance. Use algaecides or phosphate removers to prevent algae growth and nutrient buildup.

Brush and Vacuum Pool Surfaces: Brush pool walls, floors, and steps regularly to dislodge algae and prevent them from adhering to surfaces. Vacuum the pool to remove algae debris and organic matter from the water.

Shock the Pool: Shock the pool with a high dose of chlorine or non-chlorine shock treatment to kill existing algae and oxidize organic contaminants. Follow manufacturer instructions and safety precautions when using shock products.

Maintain Proper Pool Maintenance: Skim the pool surface daily to remove leaves, debris, and organic matter. Clean skimmer baskets, pump baskets, and filters regularly to ensure efficient filtration and circulation.

Brush and Treat Pool Surfaces: Use algaecide treatments to prevent algae growth and inhibit algae spores from germinating. Follow product instructions carefully and apply algaecides to pool surfaces according to dosage recommendations.

Minimize Sunlight Exposure: Use pool covers or shades to reduce direct sunlight exposure and minimize algae growth. Maintain proper water chemistry and sanitizer levels to prevent algae growth in shaded areas.

CLOUDY WATER

Cloudy water is a common issue encountered in swimming pools and can be caused by various factors affecting water clarity. Cloudiness can range from a slight haze to a completely opaque appearance, making it difficult to see the bottom of the pool. Here are some common causes of cloudy water in swimming pools:

Poor Filtration: Inadequate filtration or insufficient runtime of the pool pump can result in poor water circulation and filtration. This leads to the accumulation of debris, dirt, and other particles in the water, causing cloudiness.

Insufficient Sanitation: Low sanitizer levels, such as chlorine or bromine, allow algae, bacteria, and other microorganisms to proliferate in the pool water. This can lead to cloudy water due to microbial growth and organic contamination.

High Total Dissolved Solids (TDS): Elevated levels of total dissolved solids, which include minerals, salts, and other dissolved substances, can cause cloudiness in pool water. TDS levels may increase over time due to evaporation, chemical treatment, and water replacement.

pH Imbalance: Incorrect pH levels can affect water balance and cause cloudiness in swimming pools. Low pH (acidic) can corrode pool surfaces and equipment, while high pH (alkaline) can lead to scale formation and cloudiness.

Algae Growth: Algae blooms can result in green or yellowish-green water, contributing to cloudiness in swimming pools. Algae thrive in warm, sunlit environments with inadequate sanitation and can quickly spread if not treated promptly.

Organic Contaminants: Organic contaminants, such as leaves, grass, body oils, and sunscreen residues, can accumulate in pool water and contribute to cloudiness. These contaminants provide nutrients for microbial growth and affect water clarity.

High Calcium Hardness: Elevated calcium hardness levels can lead to scale formation on pool surfaces and equipment, contributing to cloudy water. Calcium scale may result from high levels of dissolved calcium carbonate in the water.

Poor Water Chemistry Balance: Imbalanced water chemistry, including alkalinity, calcium hardness, and cyanuric acid levels, can affect water clarity and contribute to cloudiness. Maintaining proper water balance is essential for preventing cloudiness in swimming pools.

Environmental Factors: Environmental factors such as wind-blown debris, pollen, dust, and airborne pollutants can introduce particles into the pool water, causing cloudiness. Rainwater runoff and nearby construction activities can also affect water clarity.

To address cloudy water issues in swimming pools, consider the following preventive measures and corrective actions:

Maintain Proper Filtration: Ensure that the pool pump and filter are properly sized and operated to achieve adequate water circulation and filtration. Clean or backwash the filter regularly to remove trapped debris and maintain optimal filtration efficiency.

Optimize Sanitation: Maintain proper sanitizer levels (e.g., chlorine or bromine) to control microbial growth and prevent algae blooms. Shock the pool periodically to oxidize contaminants and restore sanitizer effectiveness.

Monitor and Adjust pH Levels: Test pool water regularly and adjust pH levels to the recommended range (typically 7.2 to 7.6) using pH increaser or decreaser as needed. Proper pH balance is essential for water clarity and sanitizer effectiveness.

Brush and Vacuum Pool Surfaces: Brush pool walls, floors, and steps regularly to loosen algae, debris, and biofilm buildup. Vacuum the pool to remove suspended particles and debris from the water, especially in hard-to-reach areas.

Maintain Proper Water Chemistry Balance: Monitor alkalinity, calcium hardness, and cyanuric acid levels to ensure proper water balance and prevent scale formation. Adjust chemical levels as needed to maintain balance and prevent cloudiness.

Use Clarifiers or Flocculants: Consider using pool clarifiers or flocculants to coagulate suspended particles and improve water clarity. Clarifiers help particles clump together for easier filtration, while flocculants cause particles to sink to the bottom for vacuuming.

Remove Organic Contaminants: Skim the pool surface regularly to remove leaves, insects, and other debris. Use a pool skimmer or leaf net to capture floating debris before it sinks to the bottom and contributes to cloudiness.

Backwash the Filter: Backwash sand or DE filters and clean cartridge filters as needed to remove trapped debris and restore filtration efficiency. Follow manufacturer instructions for proper backwashing and filter maintenance procedures.

Keep Pool Area Clean: Maintain a clean pool area by removing debris, dirt, and vegetation from around the pool. Minimize the introduction of contaminants into the water to reduce the risk of cloudiness.

PH IMBALANCE

pH imbalance is a common issue encountered in swimming pools and can lead to various problems affecting water quality, swimmer comfort, and pool equipment. pH measures the acidity or alkalinity of pool water on a scale from 0 to 14, with a pH of 7 considered neutral. Here are some common pH imbalance issues in swimming pools:

Low pH (Acidic Water): Low pH levels in pool water (below 7.2) can cause several problems, including:

Skin and Eye Irritation: Acidic water can irritate the skin, eyes, and mucous membranes of swimmers, leading to discomfort and potential health concerns.

Corrosion of Pool Surfaces and Equipment: Acidic water can corrode metal fittings, plaster surfaces, and pool equipment, such as pumps, heaters, and filters.

Increased Chlorine Demand: Low pH can cause chlorine to become more reactive and less effective as a sanitizer, leading to increased chlorine consumption and difficulty maintaining proper sanitizer levels.

Etching of Pool Surfaces: Acidic water can etch and deteriorate pool surfaces, including concrete, tile, and grout, resulting in roughness and discoloration.

Staining: Low pH can contribute to metal staining, particularly in pools with metal components or copper-based algaecides.

High pH (Alkaline Water): High pH levels in pool water (above 7.8) can also lead to several issues, including:

Reduced Sanitizer Effectiveness: High pH can reduce the effectiveness of chlorine and other sanitizers, allowing algae and bacteria to proliferate and leading to cloudy water and poor water quality.

Scale Formation: Elevated pH levels can cause calcium carbonate (scaling) to precipitate out of solution and form scale deposits on pool surfaces, plumbing, and equipment.

Cloudy Water: High pH can contribute to cloudiness in pool water by promoting the precipitation of minerals and reducing the ability of the filter to remove suspended particles.

Reduced Swimmer Comfort: Alkaline water may cause skin and eye irritation and leave swimmers feeling uncomfortable and itchy.

Algae Growth: High pH can create conditions favorable for algae growth, particularly if combined with inadequate sanitation and poor water circulation.

Factors contributing to pH imbalance:

Environmental Factors: Factors such as rainfall, wind-blown debris, dust, and vegetation can introduce contaminants and affect pH levels in pool water.

Chemical Additions: Adding chemicals such as chlorine, shock treatments, algaecides, and pH adjusters (e.g., pH increaser or decreaser) can impact pH levels in the pool.

Swimmer Load: The number of swimmers and their activities in the pool can affect pH levels by introducing organic contaminants, perspiration, and body oils.

Evaporation and Water Replacement: Evaporation and water replacement can alter the concentration of dissolved minerals and affect pH levels in the pool.

> **To address pH imbalance issues in swimming pools, consider the following preventive measures and corrective actions:**

Regular Water Testing: Test pool water regularly using a reliable test kit to monitor pH levels and ensure they remain within the recommended range (typically 7.2 to 7.6 for most pools).

pH Adjustment: Adjust pH levels using pH increasers (sodium carbonate or soda ash) to raise pH or pH decreasers (sodium bisulfate or muriatic acid) to lower pH as needed. Follow manufacturer instructions and dosage recommendations carefully when adjusting pH.

Maintain Proper Alkalinity: Monitor total alkalinity levels and adjust them within the recommended range (typically 80 to 120 ppm) to help stabilize pH and prevent fluctuations.

Monitor and Adjust Total Hardness: Monitor calcium hardness levels and adjust them within the recommended range (typically 200 to 400 ppm) to prevent scale formation or etching of pool surfaces.

Preventive Maintenance: Maintain proper pool maintenance practices, including regular cleaning, brushing, vacuuming, and backwashing of filters, to minimize the introduction of contaminants and stabilize water chemistry.

Use Stabilizers: Use cyanuric acid (stabilizer) in outdoor pools to protect chlorine from degradation by sunlight and help maintain sanitizer effectiveness.

Control Swimmer Load: Limit the number of swimmers in the pool and encourage proper hygiene practices to minimize the introduction of organic contaminants and reduce the need for pH adjustments.

Address Environmental Factors: Minimize the impact of environmental factors, such as wind-blown debris, rainfall, and vegetation, by using pool covers, fencing, and landscaping to prevent contamination and maintain water quality.

CHLORINE AND SANITIZER ISSUES

Chlorine and sanitizer issues are common concerns in swimming pools and can affect water quality, swimmer safety, and overall pool maintenance. Chlorine and other sanitizers play a crucial role in killing bacteria, viruses, algae, and other microorganisms present in pool water. Here are some common chlorine and sanitizer issues encountered in swimming pools:

Inadequate Sanitizer Levels: Insufficient chlorine or sanitizer levels in pool water can lead to ineffective disinfection, allowing harmful bacteria, viruses, and algae to proliferate. This can result in poor water quality, increased risk of infections, and potential health hazards for swimmers.

Chlorine Demand: High chlorine demand refers to the amount of chlorine required to effectively sanitize pool water and oxidize organic contaminants. Factors such as heavy swimmer load, organic debris, sunlight exposure, and environmental pollutants can increase chlorine demand and deplete sanitizer levels more quickly.

Chlorine Residual: Chlorine residual refers to the concentration of free available chlorine (FAC) or combined available chlorine (CAC) remaining in pool water after disinfection. Maintaining an appropriate chlorine residual is essential for continuous sanitation and preventing microbial growth.

Chlorine Loss: Chlorine loss occurs when chlorine is consumed or degraded due to various factors, including sunlight (UV degradation), organic contaminants (chlorine demand), high temperatures, and environmental pollutants. Chlorine loss can result in rapid depletion of sanitizer levels and require frequent replenishment.

Combined Chlorine (Chloramines): Combined chlorine, also known as chloramines, forms when chlorine combines with ammonia and organic nitrogen compounds present in pool water. Chloramines produce unpleasant odors, irritate the eyes and skin of swimmers, and contribute to poor water quality and respiratory issues (e.g., "chlorine smell" or "chlorine odor").

Chlorine Lock: Chlorine lock occurs when high levels of combined chlorine (chloramines) bind up the free available chlorine (FAC), rendering it ineffective for disinfection. Chlorine lock can occur due to insufficient shock treatment, inadequate oxidation, or improper water chemistry balance.

Chlorine Stabilizer (Cyanuric Acid) Buildup: Cyanuric acid, commonly used as a chlorine stabilizer in outdoor pools, helps protect chlorine from degradation by sunlight (UV rays). However, excessive buildup of cyanuric acid can lead to "stabilizer lock" or reduced chlorine effectiveness, requiring dilution or partial water replacement.

Chlorine Irritation: High chlorine levels or chlorine spikes in pool water can cause skin and eye irritation, respiratory discomfort, and allergic reactions in swimmers. Proper management of chlorine levels is essential to ensure swimmer safety and comfort.

Chlorine Resistance: Some microorganisms, such as certain strains of bacteria or algae, may develop resistance to chlorine over time, making them more difficult to eradicate. Persistent algae growth or recurring microbial contamination may indicate the presence of chlorine-resistant organisms.

Environmental Factors: Environmental factors, including sunlight exposure, temperature fluctuations, rainfall, wind-blown debris, and swimmer load, can influence chlorine levels and sanitizer effectiveness in pool water. Regular monitoring and adjustment are necessary to counteract these influences.

To address chlorine and sanitizer issues in swimming pools, consider the following preventive measures and corrective actions:

Maintain Proper Sanitizer Levels: Monitor chlorine levels regularly using a reliable test kit and adjust them within the recommended range (typically 1 to 3 ppm for free chlorine) to ensure effective disinfection and algae prevention.

Shock the Pool: Shock the pool periodically with a high dose of chlorine or non-chlorine shock treatment to oxidize organic contaminants, eliminate chloramines, and restore sanitizer effectiveness. Follow manufacturer instructions and safety precautions when shocking the pool.

Use Algaecides: Use algaecide treatments to prevent algae growth and inhibit algae spores from germinating. Choose appropriate algaecides based on pool type, algae species, and manufacturer recommendations.

Maintain Proper Water Chemistry: Balance pH, alkalinity, calcium hardness, and cyanuric acid levels within the recommended ranges to optimize sanitizer effectiveness and prevent chemical imbalances that can affect chlorine performance.

Prevent Chlorine Loss: Minimize chlorine loss by using stabilized chlorine products (with cyanuric acid) for outdoor pools, maintaining proper water circulation and filtration, covering pools when not in use, and addressing factors contributing to chlorine demand.

Superchlorination: Implement superchlorination or breakpoint chlorination to raise chlorine levels temporarily and eliminate chloramines and organic contaminants. Follow proper procedures and safety precautions when superchlorinating the pool.

Monitor Chlorine Residual: Regularly test chlorine residual levels, including free available chlorine (FAC) and combined available chlorine (CAC), to ensure adequate disinfection and sanitation. Adjust chlorine dosing or treatment accordingly based on test results.

Maintain Proper Filtration and Circulation: Ensure proper operation and maintenance of pool filtration systems, including cleaning or backwashing filters, inspecting pump and filter components, and optimizing water circulation to enhance sanitizer distribution and effectiveness.

Minimize Environmental Factors: Minimize the impact of environmental factors such as sunlight exposure, temperature fluctuations, wind-blown debris, and rainfall by using pool covers, shade structures, and wind barriers to protect pool water and reduce chlorine loss.

Regular Maintenance and Monitoring: Implement a regular maintenance schedule for pool cleaning, water testing, equipment inspection, and chemical treatment to prevent chlorine and sanitizer issues and ensure optimal water quality and swimmer safety.

CALCIUM HARDNESS ISSUES

Calcium hardness refers to the concentration of dissolved calcium ions in swimming pool water and is an important parameter to monitor for maintaining water balance and preventing scale formation or etching of pool surfaces. Here are some common calcium hardness issues encountered in swimming pools:

Low Calcium Hardness: Low calcium hardness levels (below 200 ppm) can lead to several problems, including:

Corrosion of Pool Surfaces and Equipment: Low calcium hardness can cause the water to become aggressive and corrosive, leading to etching, pitting, or deterioration of pool surfaces, plaster, grout, and metal fittings.

Increased Metal Corrosion: Low calcium hardness can promote the leaching of metals from pool surfaces and equipment, leading to staining, discoloration, or corrosion of metal components (e.g., stainless steel, copper, aluminum).

Vinyl Liner Wrinkling or Damage: Low calcium hardness can cause vinyl pool liners to become brittle, shrink, wrinkle, or develop tears, compromising their integrity and longevity.

Water Instability: Low calcium hardness can result in unstable water chemistry and pH fluctuations, making it challenging to maintain proper water balance and sanitizer effectiveness.

High Calcium Hardness: High calcium hardness levels (above 400 ppm) can also lead to several issues, including:

Scale Formation: High calcium hardness can cause calcium carbonate (scaling) to precipitate out of solution and form scale deposits on pool surfaces, plumbing, and equipment. Scale buildup can impair water flow, reduce filtration efficiency, and damage pool components.

Cloudy Water: High calcium levels can contribute to cloudiness in pool water by promoting the precipitation of calcium carbonate and other minerals. Cloudy water affects water clarity and may require treatment to resolve.

Reduced Effectiveness of Sanitizers: High calcium hardness can interfere with the effectiveness of chlorine and other sanitizers by reducing their ability to kill bacteria, algae, and other microorganisms. This can lead to poor water quality and increased risk of infections.

Water Chemistry Imbalance: High calcium hardness can affect water balance and pH levels, making it challenging to maintain proper water chemistry. It may require additional chemical adjustments to stabilize water parameters and prevent scale formation or corrosion.

Factors contributing to calcium hardness issues:

Source Water Characteristics: The calcium hardness of source water (e.g., tap water, well water) can vary depending on regional geology and water treatment processes. Using water with high or low calcium hardness can affect pool water chemistry.

Chemical Additions: Adding calcium-based chemicals such as calcium chloride or calcium hypochlorite can increase calcium hardness levels in pool water. Overuse or improper dosing of these chemicals can lead to high calcium levels.

Water Evaporation and Dilution: Evaporation and water replacement can increase calcium hardness levels over time by concentrating dissolved minerals in pool water. In areas with high evaporation rates, regular monitoring and partial water replacement may be necessary to control calcium hardness.

pH and Alkalinity Levels: pH and alkalinity levels can influence calcium hardness by affecting the solubility of calcium carbonate in pool water. Imbalanced pH and alkalinity can lead to scale formation or calcium precipitation.

To address calcium hardness issues in swimming pools, consider the following preventive measures and corrective actions:

Regular Water Testing: Test pool water regularly using a reliable test kit to monitor calcium hardness levels and ensure they remain within the recommended range (typically 200 to 400 ppm for most pools).

Water Balancing: Adjust pH, alkalinity, and calcium hardness levels as needed to maintain proper water balance and prevent scale formation or corrosion. Use pH increasers or decreasers, alkalinity adjusters, and calcium hardness increasers or decreasers as required.

Partial Water Replacement: If calcium hardness levels are too high, consider partially draining and refilling the pool with fresh water to dilute dissolved minerals and reduce calcium hardness. Regular water replacement can help control calcium hardness buildup over time.

Calcium Hardness Reducer: Use commercial products specifically designed to lower calcium hardness levels in pool water. These products contain chelating agents or sequestering agents that bind to calcium ions and prevent them from forming scale deposits.

Scale Prevention and Treatment: Use scale inhibitors or preventatives to control scale formation and prevent calcium carbonate precipitation. These products help keep dissolved minerals in solution and inhibit scale buildup on pool surfaces, plumbing, and equipment.

Proper Filtration and Circulation: Maintain proper operation and maintenance of pool filtration systems, including cleaning or backwashing filters, to remove suspended particles and prevent scale accumulation. Optimize water circulation to distribute chemicals evenly and prevent localized scaling.

Source Water Analysis: Conduct a comprehensive analysis of source water to determine its calcium hardness and mineral content. Knowing the characteristics of source water can help anticipate potential calcium hardness issues and adjust pool water accordingly.

STAINING

Staining is a common issue encountered in swimming pools and can detract from the overall appearance of pool surfaces, affecting water clarity and swimmer satisfaction. Stains can appear in various forms, including discoloration, spots, streaks, or patches, and may be caused by different factors. Here are some common staining issues in swimming pools:

Metal Stains: Metal stains are among the most prevalent types of staining in swimming pools and can result from the presence of dissolved metals, such as iron, copper, manganese, or cobalt, in the water. Metal stains often manifest as brown, orange, green, blue, or black discoloration on pool surfaces, particularly plaster, tile, grout, or vinyl liners. Metals may enter the pool water from various sources, including source water, metal components (e.g., heaters, fittings), corrosion of pool equipment, metal-based algaecides, or environmental pollutants (e.g., well water, fertilizers).

Organic Stains: Organic stains are caused by the deposition of organic materials, such as leaves, grass, algae, or other vegetation, on pool surfaces. Organic stains may appear as green, brown, yellow, or black discoloration and can develop on pool walls, floors, steps, or in hard-to-reach areas with poor water circulation or inadequate sanitation. Algae growth, in particular, can contribute to organic staining if not treated promptly.

Calcium Scale: Calcium scale, also known as calcium carbonate or scaling, can form when calcium hardness levels are too high or when pH and alkalinity levels are imbalanced. Scale deposits often appear as white or grayish-colored crusty buildup on pool surfaces, tile lines, water features, or plumbing fixtures. High calcium hardness levels can promote scale formation, especially in areas with hard water or high evaporation rates.

Rust Stains: Rust stains occur when iron or other metals oxidize and deposit on pool surfaces, resulting in reddish-brown discoloration. Rust stains may originate from rusty pool equipment (e.g., metal fittings, heaters, ladders), iron-rich source water, or environmental contaminants (e.g., groundwater, fertilizers). Rust stains are particularly common in pools with metal components or in areas with iron-rich soil or water sources.

Salt Stains: Salt stains can develop in saltwater pools or pool environments where salt or chlorides are present. Salt stains may appear as white, crystalline deposits on pool surfaces, decking, coping, or surrounding hardscape materials. Salt deposits can accumulate due to evaporation of saltwater, overspray during pool maintenance, or splashing from wave action in saltwater pools.

Copper Stains: Copper stains can occur when copper ions oxidize and deposit on pool surfaces, resulting in blue or greenish discoloration. Copper stains may originate from copper-based algaecides, copper plumbing components, or corrosion of metal fittings or heaters. Copper staining is more common in pools with copper-based treatment products or in areas with acidic water conditions.

Environmental Stains: Environmental stains can result from external factors such as leaves, dirt, pollen, insects, bird droppings, or other airborne contaminants entering the pool water. Environmental stains may appear as brown, green, yellow, or black discoloration on pool surfaces and can be challenging to remove without proper cleaning and maintenance.

Chemical Stains: Chemical stains may occur due to improper use or overuse of pool chemicals, including chlorine, shock treatments, algaecides, or pH adjusters. Chemical stains may manifest as localized discoloration or etching of pool surfaces, particularly if chemicals are not properly diluted, balanced, or distributed in the water.

To address staining issues in swimming pools, consider the following preventive measures and corrective actions:

Regular Maintenance: Implement a regular maintenance routine, including brushing, vacuuming, skimming, and backwashing, to remove debris, prevent organic buildup, and maintain water clarity. Regular cleaning helps prevent staining and keeps pool surfaces clean and smooth.

Water Balance: Maintain proper water chemistry balance, including pH, alkalinity, calcium hardness, and sanitizer levels, to prevent scale formation, corrosion, or staining. Test pool water regularly and adjust chemical levels as needed to ensure optimal water balance.

Metal Sequestrants: Use metal sequestrants or chelating agents to bind to dissolved metals in pool water and prevent them from oxidizing and staining pool surfaces. Metal sequestrants help keep metals in solution and inhibit staining, particularly in pools with metal contamination or hard water.

Stain Removal Treatments: Use stain removal treatments or specialized stain removers designed to target specific types of stains, such as metal stains, organic stains, or scale deposits. Follow manufacturer instructions carefully and apply treatments as directed to effectively remove stains without damaging pool surfaces.

Pool Surface Protection: Apply pool surface sealants, coatings, or protectants to protect pool surfaces from staining, scaling, or corrosion. Surface treatments can help maintain the integrity and appearance of pool finishes and reduce the risk of staining or discoloration over time.

Preventative Measures: Implement preventative measures to minimize the introduction of staining agents into the pool water, such as using leaf nets, skimmer socks, or pool covers to prevent debris from entering the pool. Regularly trim vegetation, clean gutters, and maintain proper landscaping to reduce environmental contaminants.

FILTER ISSUES

Filter issues are common in swimming pools and can affect water clarity, sanitation, and overall pool maintenance. The pool filter plays a crucial role in removing debris, particles, and contaminants from the water, ensuring clean and healthy swimming conditions. Here are some common filter issues encountered in swimming pools:

Clogged or Dirty Filter: One of the most common filter issues is a clogged or dirty filter, which occurs when debris, dirt, oils, and other particles accumulate on the filter media (e.g., sand, cartridge, diatomaceous earth). A clogged filter restricts water flow, reduces filtration efficiency, and can lead to poor water quality, cloudy water, and increased pressure on the filtration system.

Filter Media Problems: Different types of pool filters (sand, cartridge, DE) may experience specific media-related issues. For example:

Sand Filter: Issues such as channeling (water bypassing filter media), clumping or channeling of sand, or ineffective backwashing can occur in sand filters.

Cartridge Filter: Cartridge filters may become clogged or damaged, leading to reduced filtration capacity and decreased water flow.

DE Filter: DE filters may experience problems such as torn or damaged filter grids, improper DE distribution, or excessive DE buildup, affecting filtration performance.

Backwashing Problems: Inadequate or improper backwashing of the filter can lead to insufficient debris removal, filter clogging, or incomplete regeneration of filter media. Backwashing too frequently or for insufficient durations may also impact filtration efficiency and water clarity.

Leaks or Cracks: Leaks or cracks in the filter tank, plumbing connections, valves, or O-rings can occur due to age, corrosion, freeze damage, or improper installation. Leaks can result in water loss, air infiltration, decreased filtration effectiveness, and potential water damage to surrounding areas.

Air Entrapment or Air Leaks: Air entrapped in the filtration system, air leaks in suction lines, or faulty seals and gaskets can cause air bubbles to appear in the pump basket or return jets. Air entrainment can reduce pump efficiency, create noise, and lead to cavitation or priming issues.

Pressure Problems: Fluctuations in filter pressure, excessively high or low pressure readings, or pressure spikes can indicate filter issues such as clogging, backwash valve problems, or pump issues. Monitoring filter pressure regularly is essential for detecting and addressing potential problems.

Broken or Damaged Components: Broken or damaged filter components, including filter housings, manifolds, grids, cartridges, or valve assemblies, can impair filtration performance and require replacement or repair. Visual inspection of filter components can identify signs of wear, corrosion, or damage.

Inadequate Filtration: Filters may not provide adequate filtration if they are undersized for the pool, improperly installed, or poorly maintained. Insufficient filtration can result in poor water quality, algae growth, and increased demand for chemical treatment.

Improper Maintenance: Neglecting regular filter maintenance tasks such as backwashing, cleaning, and replacing filter media or cartridges can lead to filter problems and reduced efficiency. Routine filter maintenance is essential for optimal performance and prolonging the lifespan of the filtration system.

To address filter issues in swimming pools, consider the following preventive measures and corrective actions:

Regular Inspection and Maintenance: Inspect the filter system regularly for signs of damage, leaks, or wear. Perform routine maintenance tasks such as backwashing, cleaning, and replacing filter media or cartridges according to manufacturer recommendations.

Proper Backwashing: Follow proper backwashing procedures as recommended by the manufacturer to ensure effective debris removal and regeneration of filter media. Backwash the filter when pressure increases by 25 to 30 percent above normal operating pressure or as needed based on water clarity.

Filter Cleaning: Clean filter elements (cartridges, grids) thoroughly using a filter cleaner or degreaser to remove accumulated debris, oils, and contaminants. Rinse cartridges or grids with water to ensure complete cleaning before reinstalling.

Regular Water Testing: Test pool water regularly to monitor water quality parameters such as pH, chlorine levels, alkalinity, and calcium hardness. Proper water chemistry balance helps reduce the workload on the filtration system and prolongs filter life.

Corrective Action: Address filter problems promptly by identifying the root cause of issues such as clogging, leaks, or pressure problems. Take corrective action, such as repairing leaks, replacing damaged components, or upgrading filter systems if necessary.

Professional Service: Seek assistance from a professional pool service technician or certified pool operator if filter issues persist or if you are unsure how to troubleshoot and resolve problems. A qualified technician can diagnose filter issues, perform repairs or replacements, and ensure proper operation of the filtration system.

Proper Sizing and Installation: Ensure the filter system is properly sized and installed according to the requirements of your pool size, flow rate, and filtration needs. Consult with a pool professional to select the appropriate filter type and size for your specific pool application.

Preventative Maintenance: Implement preventative maintenance practices such as regular cleaning, lubrication of seals and O-rings, and inspection of filter components to prevent premature wear, corrosion, or damage. Proactive maintenance helps prolong the lifespan of the filtration system and ensures consistent water quality.

LEAKAGE ISSUES

Leakage is a common issue encountered in swimming pools and can lead to water loss, structural damage, and increased maintenance costs if left unresolved. Identifying and addressing leaks promptly is essential for maintaining the integrity and functionality of the pool. Here are some common leakage issues in swimming pools:

Structural Leaks: Structural leaks occur when there is damage or deterioration to the pool's structural components, such as the shell, concrete, fiberglass, or vinyl liner. Structural leaks can result from cracks, fractures, shifting soil, settlement, or improper construction techniques. Structural leaks may manifest as visible cracks in the pool shell, bulging or shifting of pool walls, or water loss from the pool.

Plumbing Leaks: Plumbing leaks occur in the pool's circulation system, including pipes, fittings, valves, and connections. Plumbing leaks can result from corrosion, wear and tear, freeze damage, ground movement, or poor installation. Plumbing leaks may manifest as damp or wet areas around pipes or equipment, loss of water pressure, air bubbles in the pump basket, or water seepage around fittings.

Equipment Leaks: Equipment leaks occur in pool components such as pumps, filters, heaters, chlorinators, or skimmers. Equipment leaks can occur due to worn seals, damaged gaskets, cracked housings, or faulty connections. Equipment leaks may manifest as drips, puddles, or wet spots around equipment, visible cracks or damage to equipment components, or water loss during operation.

Structural Movement: Structural movement, including settlement, subsidence, or ground shifting, can cause stress on the pool shell and plumbing, leading to leaks. Soil erosion, expansive soils, tree roots, seismic activity, or nearby construction can contribute to structural movement and potential leakage issues. Structural movement may manifest as uneven pool surfaces, gaps between coping and decking, or cracks in surrounding structures.

Expansion Joint Failure: Expansion joints are installed between pool components (e.g., coping, deck, shell) to accommodate thermal expansion and contraction and prevent cracking. Failure of expansion joints due to deterioration, displacement, or improper installation can lead to water infiltration and leakage. Expansion joint failure may manifest as gaps, separation, or displacement of pool components.

Vinyl Liner Leaks: Vinyl liner leaks occur in pools with vinyl liners due to tears, punctures, or deterioration of the liner material. Vinyl liner leaks can result from sharp objects, improper installation, age-related wear, or chemical imbalances. Vinyl liner leaks may manifest as visible tears, wrinkles, or bulges in the liner, loss of water level, or wet spots around the pool perimeter.

Skimmer or Drain Leaks: Leaks can occur in pool skimmers, drains, or fittings due to damage, deterioration, or improper installation. Skimmer or drain leaks may result from cracked or broken components, loose fittings, or faulty seals. Skimmer or drain leaks may manifest as water loss around the skimmer or drain area, reduced suction, or air bubbles in the skimmer basket.

Surface Leaks: Surface leaks occur when water seeps through cracks, gaps, or imperfections in the pool's surface material, such as plaster, tile, or fiberglass. Surface leaks can result from age-related wear, settlement, or improper installation. Surface leaks may manifest as dampness, staining, or efflorescence on pool surfaces, or water loss from the pool.

Groundwater Seepage: Groundwater seepage occurs when water infiltrates the pool structure from the surrounding soil or water table. Groundwater seepage can occur in pools located in areas with high water tables, poor drainage, or hydrostatic pressure issues. Groundwater seepage may manifest as rising water levels in the pool, moisture or dampness in the pool structure, or water intrusion during heavy rainfall or flooding.

To address leakage issues in swimming pools, consider the following preventive measures and corrective actions:

Leak Detection: Conduct a comprehensive leak detection assessment using specialized equipment, such as leak detection devices, pressure testing, dye testing, or acoustic listening devices, to pinpoint the location and extent of leaks in the pool structure, plumbing, or equipment.

Repair or Replacement: Repair or replace damaged or deteriorated pool components, including cracked shells, damaged plumbing, worn seals, or faulty equipment. Seek assistance from a professional pool service technician or certified pool operator to perform repairs or replacements as needed.

Pressure Testing: Perform pressure testing of the pool's plumbing system to identify leaks, air leaks, or pressure fluctuations. Pressure testing helps locate hidden leaks in pipes, fittings, or valves and ensures the integrity of the circulation system.

Surface Repair: Repair cracks, gaps, or imperfections in the pool's surface material using appropriate patching compounds, sealants, or coatings. Surface repair helps prevent water infiltration and restores the structural integrity of the pool's finish.

Expansion Joint Maintenance: Inspect and maintain expansion joints regularly to ensure they remain watertight and flexible. Replace deteriorated or damaged expansion joint material and reseal joints as needed to prevent water infiltration and structural movement.

Vinyl Liner Inspection: Inspect vinyl liners for tears, punctures, or damage regularly and repair or replace liners as needed. Ensure proper installation and maintenance of vinyl liners to prevent water leakage and extend their lifespan.

EQUIPMENT MALFUNCTION

Equipment malfunction issues in swimming pools can disrupt pool operation, compromise water quality, and affect the overall swimming experience. Here are some common equipment malfunction issues encountered in swimming pools:

Pump Problems: Pump failure or malfunctions can occur due to various reasons such as motor issues, electrical problems, clogged impellers, or worn-out seals. Symptoms of pump problems include reduced water circulation, low flow rate, unusual noises, or pump motor overheating.

Filter Issues: Filter problems, including clogging, leaks, damaged filter media, or valve malfunctions, can impact filtration efficiency and water clarity. Common symptoms of filter issues include high pressure readings, cloudy water, poor water circulation, or debris bypassing the filter.

Heater Malfunctions: Heater problems, such as ignition failure, sensor issues, thermostat malfunction, or scale buildup, can prevent the pool water from reaching the desired temperature. Signs of heater malfunctions include inconsistent heating, no heat output, error codes, or abnormal noises.

Chlorinator or Salt System Failures: Chlorinator or salt system failures can lead to inadequate sanitizer levels, algae growth, or waterborne contaminants. Issues such as clogged cells, malfunctioning control panels, or salt system errors can affect sanitizer production and water quality.

Automatic Cleaner Problems: Automatic pool cleaners may experience issues such as tangled hoses, stuck wheels, jammed gears, or motor malfunctions, affecting their cleaning performance. Signs of automatic cleaner problems include incomplete cleaning, erratic movement, or failure to operate.

Timer or Control Panel Malfunctions: Timer or control panel malfunctions can disrupt the scheduling and operation of pool equipment, including pumps, heaters, lights, or automated features. Symptoms of timer or control panel issues include incorrect time settings, unresponsive controls, or error messages.

Lighting Failures: Lighting failures, such as burned-out bulbs, faulty wiring, or corroded fixtures, can affect pool visibility, safety, and aesthetics. Signs of lighting problems include flickering lights, dim illumination, or complete light failure.

Valve Leaks or Sticking: Valve leaks or sticking can occur in diverter valves, check valves, or multiport valves, leading to water loss, improper water flow direction, or difficulty in adjusting settings. Symptoms of valve problems include water drips, pool water loss, or difficulty in switching between modes.

Leak Detection: Conduct a comprehensive leak detection assessment using specialized equipment, such as leak detection devices, pressure testing, dye testing, or acoustic listening devices, to pinpoint the location and extent of leaks in the pool structure, plumbing, or equipment.

Repair or Replacement: Repair or replace damaged or deteriorated pool components, including cracked shells, damaged plumbing, worn seals, or faulty equipment. Seek assistance from a professional pool service technician or certified pool operator to perform repairs or replacements as needed.

Pressure Testing: Perform pressure testing of the pool's plumbing system to identify leaks, air leaks, or pressure fluctuations. Pressure testing helps locate hidden leaks in pipes, fittings, or valves and ensures the integrity of the circulation system.

Surface Repair: Repair cracks, gaps, or imperfections in the pool's surface material using appropriate patching compounds, sealants, or coatings. Surface repair helps prevent water infiltration and restores the structural integrity of the pool's finish.

Expansion Joint Maintenance: Inspect and maintain expansion joints regularly to ensure they remain watertight and flexible. Replace deteriorated or damaged expansion joint material and reseal joints as needed to prevent water infiltration and structural movement.

Vinyl Liner Inspection: Inspect vinyl liners for tears, punctures, or damage regularly and repair or replace liners as needed. Ensure proper installation and maintenance of vinyl liners to prevent water leakage and extend their lifespan.

Skimmer and Drain Maintenance: Inspect skimmers, drains, and associated fittings regularly for signs of damage, leaks, or blockages. Clean skimmer baskets, drain covers, and strainer baskets to prevent debris buildup and maintain proper water flow.

Repairing equipment malfunctions in a swimming pool typically requires careful diagnosis of the issue followed by appropriate corrective actions. Here's a general guide on how to repair common equipment malfunctions:

Diagnosis: Identify the specific equipment malfunction by observing symptoms, conducting visual inspections, and testing equipment operation. Determine whether the issue lies with the pump, filter, heater, chlorinator, cleaner, timer/control panel, lighting, valves, or other components.

Gather Tools and Supplies: Collect the necessary tools and supplies for the repair, including wrenches, screwdrivers, pliers, pipe cutters, replacement parts, lubricants, sealants, and cleaning materials. Ensure that you have the correct tools for the type of equipment and repair task at hand.

Safety Precautions: Prioritize safety by turning off power to the equipment, shutting off relevant valves, and following manufacturer instructions and safety guidelines. Wear appropriate personal protective equipment (PPE) such as gloves, goggles, and insulated tools when working with electrical components or chemicals.

Pump Repair: For pump issues such as motor failure, clogged impellers, or leaking seals, follow these steps:

- Disconnect power to the pump and drain water from the

pump basket.

- Remove the pump housing or cover to access internal components.

- Inspect the motor, impeller, seals, and O-rings for damage or wear.

- Replace faulty components, clean debris, and lubricate moving parts as needed.

- Reassemble the pump, ensuring proper alignment and tight connections.

- Test the pump for proper operation and water flow.

Filter Maintenance: For filter problems such as clogging, leaks, or damaged media, perform the following tasks:

- Turn off the pump and relieve pressure from the filter.

- Open the filter housing or access panel to inspect the filter media, cartridges, or grids.

- Clean or replace dirty or damaged filter media, cartridges, or grids.

- Check for leaks in the filter housing, valves, or plumbing connections.

- Repair or replace faulty components, gaskets, or seals as necessary.

- Reassemble the filter, ensuring proper seating and secure connections.

- Restart the pump and monitor filtration performance and pressure.

Heater Troubleshooting: For heater malfunctions such as ignition failure, sensor errors, or scale buildup, follow these steps:

- Turn off power to the heater and allow it to cool down.

- Inspect the heater components, including burners, ignition system, sensors, and the heat exchanger.

- Clean or replace clogged burners, ignition components, or sensors.

- Descale the heat exchanger using appropriate descaling agents or procedures.

- Check for gas leaks, faulty wiring, or electrical issues.

- Test the heater operation and monitor temperature rise and gas combustion.

Chlorinator or Salt System Repair: For chlorinator or salt system failures, such as low sanitizer levels or malfunctioning cells, take the following actions:

- Check the chlorinator or salt system control panel for error messages or indicators.

- Inspect the cell or electrodes for buildup, scaling, or corrosion.

- Clean or replace the cell, electrodes, or sensors as needed.

- Verify proper water flow, salt levels, and system settings.

- Calibrate the system or adjust output settings as necessary.

- Monitor sanitizer levels and system operation for consistent performance.

Automatic Cleaner Maintenance: For automatic cleaner issues, such as tangled hoses, jammed gears, or motor failures, perform the following tasks:

- Disconnect power to the cleaner and remove it from the pool.

- Inspect the cleaner for debris, obstructions, or tangled parts.

- Clean or replace worn-out brushes, wheels, or gears.

- Lubricate moving parts and seals to ensure smooth operation.

- Reassemble the cleaner and test it in the pool for proper movement and cleaning.

Timer or Control Panel Troubleshooting: For timer or control panel malfunctions, such as incorrect settings or unresponsive controls, take these steps:

- Verify power supply and electrical connections to the timer or control panel.

- Check for blown fuses, tripped breakers, or loose wiring.

- Reset the timer or control panel to factory settings and reprogram as needed.

- Inspect buttons, switches, or displays for damage or wear.

- Replace faulty components, circuit boards, or control panels if necessary.

- Test the timer or control panel functions and ensure proper operation.

Lighting Repair: For lighting failures such as burned-out bulbs, faulty wiring, or corroded fixtures, follow these steps:

- Turn off power to the lighting system and drain water from light niches.

- Remove the light fixture or cover to access internal components.

- Inspect bulbs, sockets, wiring, and connections for damage or corrosion.

- Replace burned-out bulbs, damaged wiring, or faulty fixtures.

- Clean or replace lens covers, gaskets, or seals to prevent water intrusion.

- Reassemble the lighting system, ensuring watertight seals and secure connections.

- Test the lights underwater for proper illumination and waterproofing.

Valve Inspection and Repair: For valve leaks or sticking, perform the following tasks:

- Shut off water flow to the affected valve and relieve pressure from the system.

- Inspect the valve body, seals, gaskets, and moving parts for damage or wear.

- Clean debris, corrosion, or mineral deposits from valve components.

- Lubricate valve stems, O-rings, or seals with compatible lubricants.

- Repair or replace damaged or worn-out valve components as needed.

- Reassemble the valve, ensuring proper alignment and sealing.

- Test the valve operation and monitor for leaks or sticking.

Safety Checks and Testing: After completing the repairs, perform safety checks and functional testing to ensure proper operation, water flow, and equipment performance. Monitor equipment parameters, pressure readings, temperature levels, and water chemistry to verify that the repairs have resolved the issues.

LACK OF PROPER CIRCULATION

Lack of proper circulation in a swimming pool can lead to various water quality issues, including poor distribution of chemicals, uneven heating, and stagnation of water. Here are some common reasons for inadequate circulation in swimming pools and ways to address them:

Improperly Sized or Positioned Return Inlets: If the return inlets are not properly sized or positioned, they may not effectively distribute water throughout the pool, leading to areas with inadequate circulation. Ensure that the return inlets are positioned strategically to promote proper water movement and circulation. Consider adjusting the angle or direction of the return jets to improve circulation patterns.

Blocked or Obstructed Return Inlets: Blocked or obstructed return inlets can restrict water flow and hinder circulation in the pool. Inspect the return inlets for debris, vegetation, or other obstructions that may impede water movement. Remove any blockages and ensure that the return inlets are clear and unobstructed to promote optimal circulation.

Insufficient Pump Size or Flow Rate: If the pool pump is undersized or the flow rate is insufficient, it may struggle to circulate water effectively throughout the pool. Consider upgrading to a more powerful pump or adjusting the pump's speed settings to increase flow rates and improve circulation. Consult with a pool professional to determine the appropriate pump size and flow rate for your pool.

Clogged or Dirty Filter: A clogged or dirty filter can restrict water flow and hinder circulation in the pool. Regularly clean or backwash the filter to remove debris, dirt, and other contaminants that may accumulate and impede filtration and circulation. Replace filter media or cartridges as needed to maintain optimal filtration efficiency.

Blocked or Restricted Skimmer Baskets: Blocked or restricted skimmer baskets can impede water flow and reduce circulation in the pool. Inspect the skimmer baskets for debris, leaves, or other obstructions that may restrict water intake. Clean or empty the skimmer baskets regularly to ensure unrestricted water flow and promote efficient circulation.

Inadequate Number of Skimmer Baskets: If the pool has an inadequate number of skimmer baskets, it may not be able to effectively remove surface debris and promote proper circulation. Consider adding additional skimmer baskets to improve surface skimming and water intake, especially in larger or heavily used pools.

Inadequate or Improperly Positioned Pool Returns: Inadequate or improperly positioned pool returns can contribute to poor circulation and dead spots in the pool. Evaluate the placement and number of pool returns to ensure even distribution of water throughout the pool. Consider adding additional pool returns or adjusting existing returns to optimize circulation patterns.

Blocked or Restricted Suction Lines: Blocked or restricted suction lines can diminish water flow and impede circulation in the pool. Inspect the suction lines for debris, leaves, or other obstructions that may obstruct water intake. Clear any blockages and ensure that the suction lines are free from restrictions to promote efficient circulation.

Lack of Proper Water Chemistry Balance: Imbalanced water chemistry, such as improper pH, alkalinity, or sanitizer levels, can affect water density and hinder circulation in the pool. Maintain proper water chemistry balance by regularly testing and adjusting pH, alkalinity, sanitizer levels, and other parameters as needed. Balanced water chemistry promotes optimal circulation and prevents scale formation or corrosion in the pool.

Broken or Malfunctioning Pool Equipment: Broken or malfunctioning pool equipment, such as pumps, filters, valves, or automation systems, can disrupt circulation and hinder water movement in the pool. Inspect pool equipment for signs of damage, wear, or malfunction. Repair or replace faulty equipment components to restore proper operation and circulation in the pool.

Repairing issues related to lack of proper circulation in a swimming pool requires identifying the root cause of the problem and taking appropriate corrective actions. Here's how to repair common circulation issues:

Inspect and Clean Return Inlets: Remove any debris or obstructions blocking the return inlets. Use a brush or vacuum to clean the inlets thoroughly. Ensure that the return inlets are properly positioned and angled to promote optimal water circulation.

Check Pump and Filter: Inspect the pool pump and filter for any signs of damage, clogging, or wear. Clean or replace the filter media if necessary. Ensure that the pump is functioning correctly and that there are no leaks in the system.

Adjust Pump Speed and Run Time: Adjust the pump speed and run time to increase water flow and circulation. Higher pump speeds and longer run times can help improve circulation in the pool.

Clean Skimmer Baskets and Suction Lines: Clean skimmer baskets and suction lines to remove any debris or blockages that may be restricting water flow. Ensure that the skimmer baskets are installed correctly and that there are no leaks in the suction lines.

Balance Water Chemistry: Test the pool water chemistry and adjust pH, alkalinity, and sanitizer levels as needed. Balanced water chemistry helps prevent scale buildup and corrosion, which can hinder circulation.

Inspect Pool Returns: Check the pool returns for proper positioning and alignment. Adjust the direction of the returns to ensure even water distribution throughout the pool.

Check Valves and Fittings: Inspect valves and fittings for leaks, cracks, or damage. Repair or replace any faulty valves or fittings to restore proper circulation.

Consider Adding Additional Equipment: If necessary, consider adding additional equipment such as booster pumps, circulation pumps, or skimmer baskets to improve water circulation in problem areas of the pool.

Monitor and Test: After making repairs and adjustments, monitor the pool's circulation system regularly and conduct tests to ensure that water circulation is optimized. Adjust settings or make further repairs as needed.

Repairing issues related to environmental factors in a swimming pool involves a combination of preventative measures and targeted interventions to address specific problems. Here's how to repair common environmental issues in a swimming pool:

Algae Growth: Shock the pool with a high dose of chlorine to kill existing algae. Brush the pool walls and floor to remove algae buildup. Add an algaecide to prevent future algae growth. Maintain proper water balance and filtration to prevent algae recurrence.

Disinfectant Degradation: Regularly test and adjust chlorine levels to ensure adequate sanitization. Use stabilized chlorine products to protect against UV degradation. Consider installing a UV or ozone sanitation system to supplement chlorine disinfection.

Chemical Imbalance: Test pool water regularly and adjust pH, alkalinity, and sanitizer levels as needed. Use pH increasers or decreasers to maintain proper pH balance. Shock the pool with chlorine or nonchlorine shock to address chemical imbalances.

Debris Accumulation: Skim the pool surface and vacuum the pool regularly to remove debris. Use a pool cover when the pool is not in use to prevent debris accumulation. Install a leaf skimmer or surface cleaner to automate debris removal.

Rainwater Contamination: Shock the pool after heavy rainfall to restore chlorine levels. Test and adjust water chemistry to counteract dilution from rainwater. Maintain proper water circulation and filtration to remove contaminants.

Air Pollution: Use a pool cover when the pool is not in use to prevent airborne pollutants from entering the water. Install air filtration systems or landscaping barriers to minimize air pollution around the pool area.

Vegetation and Landscaping: Trim trees and shrubs to minimize leaf and debris shedding into the pool. Install a mesh or solid pool cover to prevent debris accumulation during landscaping maintenance.

Wildlife and Insects: Use bird deterrents, such as scare devices or netting, to discourage birds from landing near the pool. Keep pool covers securely fastened when not in use to prevent wildlife access. Use insect repellents or traps to control insect populations around the pool area.

Humidity and Condensation: Improve ventilation around the pool area to reduce humidity levels. Use dehumidifiers or air conditioning to control indoor humidity levels. Regularly clean and disinfect pool surfaces and equipment to prevent mold and mildew growth.

Groundwater and Soil Composition: Consult with a pool professional to assess and address groundwater issues affecting the pool structure. Implement drainage systems or waterproofing measures to mitigate groundwater infiltration.

Regular Maintenance: Implement a regular maintenance schedule for pool cleaning, water testing, and equipment inspection. Address any issues promptly to prevent them from escalating into larger problems.

9 798822 963009